Sinagua Sunwatchers

An Archaeoastronomy Survey of the Sacred Mountain Basin

REVISED AND EXPANDED

Kenneth J. Zoll

D1132214

VVAC Press
Camp Verde

Published by VVAC Press, a division of the Verde Valley Archaeology Center
385 South Main Street, Camp Verde, Arizona 86322-7272
928-567-0066
E-mail: vvacpress@verdevalleyarchaeology.org
www.verdevalleyarchaeology.org

All photos by author unless otherwise attributed.
Visit the companion website at www.sinaguasunwatchers.com

ISBN: 0982037805
ISBN-13: 978-0982037805

Library of Congress Control Number: 2008907400

CONTENTS

ACKNOWLEDGMENTS

This multi-year survey at the V Bar V Heritage Site could not have taken place without the encouragement and support of Peter J. Pilles, Jr., Forest Archaeologist for the Coconino National Forest. Even though he was "skeptical" at first, it was offered in a constructive way. His challenge to me that only a full year of observation could remove some of that skepticism was at first daunting, but little did he or I realize at the time what an amazing site this would turn out to be.

Of course none of this would have been possible without the opportunity that the Friends of the Forest Sedona provided me to serve as a volunteer docent at the heritage sites. This organization's dedication to the protection and interpretation of these heritage sites is an inspiring and valuable contribution to the U.S. Forest Service and to the residents of, and visitors to, the Sedona area. Thanks to Dean Campbell who provided me with excellent docent training and for his comments on the original edition.

Thanks first goes to my wife Nancy for her patience and support during the many "photo days." Many thanks for their support and assistance to Jim Graceffa, Dr. David Wilcox at the Museum of Northern Arizona, Lawrence Wasserman of Lowell Observatory, Louis Klitzke retired professor at the University of Wisconsin-Stout, and to Ekkehart Malotki, professor emeritus at Northern Arizona University.

Special thanks to Dr. Todd Bostwick who provided early inspiration in archaeoastronomy and later training, and to geologist Paul Lindberg for his insightful observations and continued encouragement. The subsequent studies during 2012 at the site could not have even been envisioned without the enthusiastic participation of these two professionals.

SINAGUA SUNWATCHERS

1

INTRODUCTION

Throughout history, the ability of a people to survive and thrive has been tied to environmental conditions. The skill to predict the climatic change of the seasons was an essential element in the ability to "control" those conditions. Seasonal calendars thus became the foundation of early cultures: hunting and gathering, planting and harvesting, worshiping and celebrating were activities dictated by specific times of the year. All of these activities have fostered the identity and strength of cultures.

The prehistoric occupants of the northern Southwest consisted of regionally identified groups of people who shared various elements of a generalized Puebloan life style. One of these groups, who lived in the Verde Valley and around the San Francisco Peaks, has been named the **Sinagua** by archaeologists. What permitted these cultures to thrive and expand was their ability to develop farming techniques that were adapted to the harsh environment. The ability to predict the proper time to plant their crops in a marginal climatic environment would have been an essential aid for their success. In addition, based on the practices of historic Pueblo people, the scheduling of various religious ceremonies (whether related to planting or not) is likely to have been an equally important calendrical requirement.

The Sinagua would have come to realize that all the events that they could observe in the sky were happening in a regular and repetitive manner. They could see the recurring path of the sun throughout the year, the repetitive changing shape of the moon and the repetitive motion and pattern of stars. They would also have noted the warming provided by the sun at certain times of the year, and the cooling effect as the sun appeared to move to the south. With regular observations, they could fashion rudimentary calendars to track and predict these events.

The goal of **archaeoastronomy** is to understand how these early skywatchers fashioned and refined systems for regulating their calendars around celestial events, both cyclical and unique. Another term that is sometimes used is cultural astronomy. This term describes the study of the diverse ways in which cultures perceived and integrated the objects in the sky into their worldview.

Archaeoastronomy, or cultural astronomy, draws on the disciplines of astronomy, archaeology and ethnology to identify and interpret these ancient markings and alignments. As will be described, ancient cultures found a variety of ways to chart the arrival of the equinoxes and solstices. These included observing the sun's position on the horizon at sunrise/sunset, patterns of sunlight and shadow on pictographs or petroglyphs, and the alignment of the sun or moon with architectural structures.

Archaeoastronomy is one of the disciplines that have appeared in recent years as a way of "interpreting" rock art. Dozens of interactions have been claimed for sites around the world, and especially here in the Southwest. There is no question that astronomical observations were important to cultures throughout the world, at all stages of cultural development. But it is difficult to evaluate claims of deliberate, planned interactions with all the variables that exist in astronomical cycles, let alone the cultural variables that we can never know with certainty. So the first question enquiring minds have to ask is "So there's an interesting interaction on the summer solstice – what happens during the rest of the year?" As far as can be determined through an extensive search of available research, proponents of archaeoastronomical explanations of rock art have seldom done such an extensive test. This gap in archaeoastronomy studies is what prompted the extensive documenting of this site. It is believed that this is the major value of the survey - it provides a complete year of data that everyone can now look at to evaluate claims of potential astronomical interactions at the V Bar V Heritage Site, with its extensive petroglyph panels.

This book was originally published as *Sinagua Sunwatchers: An Archaeoastronomy Survey of the V-V Heritage Site*. The Second Edition included the discovery of the sun shrines from the Sinagua sunwatcher position in the Sacred Mountain Basin. This discovery confirmed that the Sinagua sunwatchers used horizon watching to determine the time of year. In addition, that edition was prompted by comments of many readers who suggested that color photos of the effects would be more meaningful. This Third Edition includes a chapter on the 2011 scaffold project that provided additional data on the site.

2

V BAR V HERITAGE SITE

SITE FEATURES

The V Bar V Heritage Site is located about 12 miles southeast of Sedona, Arizona, within the Red Rock District of the Coconino National Forest. The site was originally discovered soon after Euro-American ranchers settled the Beaver Creek area in the 1870's. Although known to archaeologists since the early 1900's, its significance was not known until the U.S. Forest Service acquired the V Bar V Ranch and several parcels of land along Beaver Creek as part of a land exchange in 1994. Preliminary studies indicated that the V Bar V Heritage Site was the largest petroglyph site in the Verde Valley, and one of the best preserved. In order to protect the site and make it available to the public, the Forest initiated a series of volunteer projects and documented the site between 1994 and 1996. Through a partnership with the Friends of the Forest Sedona, the site was opened to the public in 1996.

It has long been noticed that several petroglyphs extended below the modern ground surface. In 2005, a limited excavation project was conducted in front of the panels. Several occupation surfaces were found with pottery sherds, stone flakes, grinding tools and a mano cache, indicating use of the site since A.D. 800 and likely earlier.

A sandstone bluff, heavily coated in desert varnish contains over 1,030 petroglyphs (Figure 1). The excavations also confirmed the existence of petroglyphs below the current ground level, however, their full extent must await later excavations. Most petroglyphs are on west-facing panels, although two panels face north. The astronomical images described in this study are on one of the primary west-facing panels, which will be referred to as the *Solar Panel.* Of the more than 125 petroglyph elements on this panel, eleven are thought to have astronomical associations. As can be seen from Table 1, the Solar Panel is nearly vertical and very close in alignment to true north.

Figure 1. The V Bar V Solar Panel with astronomical images

Table 1. V Bar V Site Identification

Elevation:	3,695 Ft. (1,126m)
Latitude:	North 34° 39.6'
Longitude:	West 111° 43.1'
Magnetic Declination:	11° 21' East
Solar Panel:	8° West of North
	3° Vertical declination East

Wet Beaver Creek, and the other perennial waterways of the Verde River, provide an oasis that has long attracted people to the Verde Valley. The earliest Paleo-Indian hunters occupied the Valley around 11,500 B.C. based upon Clovis-style projectile points that have been found in the area. They gave way to Archaic Period hunters and gatherers until about A.D. 600 when a change to a sedentary agricultural society becomes evident. By using a variety of both dry farming techniques as well as irrigation, these early farmers grew crops of corn, beans, several varieties of squashes, and cotton. Although agriculture was important, the wild plant and animal resources of the Verde Valley continued to provide an important part of their diet.

Archaeologists have named these prehistoric farming people the Sinagua, and have distinguished two groups of them: those who lived around the San Francisco Peaks in villages such as Elden Pueblo, Walnut Canyon, and Wupatki National Monuments, are referred to as the Northern Sinagua, while those who lived in the Verde Valley in pueblos such as Sacred Mountain, Montezuma Castle, and Tuzigoot, are known as the Southern Sinagua.

Between A.D. 600 and 1400, the Southern Sinagua developed a thriving economy and was actively involved in trading pottery, cotton cloth, and probably food to other people in northern Arizona. By 1300, they lived in pueblos that were evenly spaced at 1.8 mi. intervals along the Verde River and its tributaries (Pilles 1981). Numerous pueblo villages occur in the Wet Beaver Creek drainage. It was believed that the V Bar V Heritage Site location was used by the Southern Sinagua between A.D. 1150 and 1400 (Pilles 1996b), although excavations in 2005 suggest the site may have been used as early as A.D. 600 when the Sinagua first entered this Verde Valley (Pilles personal communication).

High above the flood plain of Wet Beaver Creek is one of the best-preserved cliff dwellings in North America. The Montezuma Castle National Monument is a five-story, 20-room cliff dwelling that served as a "high-rise apartment building" for the Southern Sinagua from about A.D. 1000 to 1400. Early settlers to the area incorrectly assumed that the imposing structure was associated with the Aztec emperor Montezuma.

Further up the creek, in what today is Lake Montezuma Estates, is a hilltop structure. What is unique about this structure is that it contained two community rooms. Most large sites of this phase have only one such room. Walls at this dwelling are over 6-½ feet (2 meters) high.

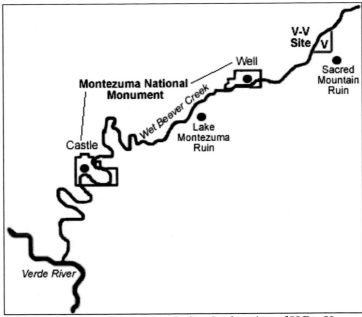

Figure 2. Wet Beaver Creek showing location of V Bar V
and other major Sinagua sites

About eleven miles (17.7 km) up Wet Beaver Creek from the Castle is another set of dwellings surrounding Montezuma Well, also part of the Montezuma Castle National Monument. This limestone sinkhole formed by the collapse of an immense underground cavern. Over one and a half million gallons of water a day flow continuously, providing a lush oasis in the midst of surrounding desert grassland. The Well empties into Wet Beaver Creek. An irrigation ditch, dug by the Sinagua to irrigate their fields with water from the Well, can still be seen at the Montezuma Well picnic area.

Wet Beaver Creek continues northeast with long sections surrounded by high walls of sandstone. About three miles (4.8 km) northeast of the Well, between two natural access points to the creek, is the V Bar V Heritage Site. Clearly visible to any traveler along the creek are the desert varnished red bluffs of the V Bar V.

There is no evidence of any large dwelling at the V Bar V petroglyph covered bluffs, although three small dwellings were recently recorded on the ridge above the petroglyphs, and a pithouse village exists within the Heritage Site boundaries.. However, a little more than a half-mile (1 km) east of the V Bar V site is an isolated white-limestone mesa known as Sacred Mountain. It contains the remains of a 60-room pueblo around a central plaza. It is thought to be the pre-eminent village of what is known as the Beaver Creek Community. The area surrounding the ruin is known as the **Sacred Mountain Basin** and may have been the largest agricultural production area in the middle Verde Valley (Fish and Fish, 1984).

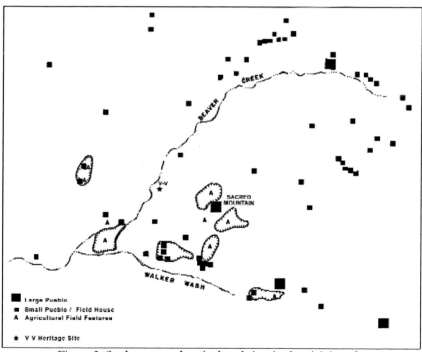

Figure 3. Settlement and agricultural sites in the vicinity of
Sacred Mountain and the V Bar V Heritage Site

As described by Fish and Fish (1984), the Sacred Mountain Basin field area for the community consists of (Figure 3) an extensive agriculture system and land use pattern, bordered by Wet Beaver Creek (west), Walker Wash (south) and rocky hills (west and north). A total of 131,500 square yards (110,000 sq m) are covered by agricultural alignments, with 77 percent of the area arranged in cobble-outlined (waffle) plots. Irrigation was enhanced by 7,600 feet (2,320 m) of canals. These canals irrigated a total of 129,200 square yards (108,000 sq m) of the field system.

Fish concluded that based on the prehistoric conditions, "the Beaver Creek fields must have been optimal farming areas, with fields simultaneously watered and enriched by irrigation." Pollen analysis indicated both corn and cotton as crops in these fields. With an average growing season of 191 days at the Sacred Mountain fields (63-100 days required for corn), a double cropping was possible but not probable. Two or three timed plantings were more likely. As a result, knowing the optimal time for seeding would have been crucial for a successful harvest.

Another feature of significance is the presence of a ballcourt at the base of Sacred Mountain. Wilcox (1991) has classified this ballcourt as an "early classic period Hohokam court" ca. A.D. 1075-1250. It is of rock construction and measures 105 ft. (32 m) in length and 78 ft. (23.8 m) in width. These courts are associated with ballgames that originated in Mesoamerica (defined as Mexico and Central America). There are over 200 documented courts in Arizona that Wilcox holds are indications that southwestern cultures evolved while interacting with the great Mesoamerican traditions. The significance of this feature, associated with the V Bar V petroglyphs, will be explored later in this study.

ETHNOGRAPHIC FRAMEWORK

There is evidence to support the claim that light and shadows react with rock art images at certain sites in the Southwest, as will be described for the V Bar V site. The evaluation of such phenomenon for astronomical significance must, however, be within the context of what we know about the people of the prehistoric Southwest (Young 1986). The use of ethnographic analogy of contemporary Puebloan ceremonial and agricultural practices is a valid tool for such evaluation. Adaptive strategies of the historic Pueblos for accurate calendars provide a reasonable way to generate a hypothesis about the astronomical practices (Zeilik 1986) of the prehistoric period that will be described for the V Bar V site.

Establishing a Sinagua cultural connection to any archaeoastronomy in the Verde Valley must, therefore, start with the ethnoastronomy of the historic Pueblos. Hopi traditions about these and other Sinagua sites have tended to be supported by archaeological evidence. Archaeologist Jesse Walter Fewkes of the Smithsonian Institution came to the Verde Valley in 1895 to conduct a survey of the ruins near the headwaters of the Verde River and the upper valley, north of Camp Verde to the area around Sedona. He was principally concerned with the survey and scientific analysis of the prehistoric resources of the region. Fewkes concentrated his study on the cliff dwellings around Oak Creek Canyon. His report (Fewkes 1898) includes a rather detailed geological, archaeological and cultural description of Montezuma Well and its ruins. In addition, he commented on the Hopi people's familiarity with the Well and the references to the site in their mythology.

The Hopi have specific traditions about the Verde Valley and various pueblo ruins that relate to their migration traditions. In the Sacred Mountain Basin land use study (Fish and Fish 1984), field boundary markers were noted which are similar to those found in historic Hopi and Zuni fields. Detailed evaluation of data provided by archaeology, ethnology, oral history, physical anthropology, architecture, technology, and sociology has resulted in a formal determination that the Hopi are the most closely affiliated modern group to the culture and lifeways of the prehistoric Sinagua (Pilles 1996a).

McCluskey (1977: 175) believed that "the Hopi provide an ideal opportunity to study the actual operation of a system of pre-scientific astronomy." He noted that the Hopi villages remained relatively uninfluenced by European missionaries or settlers until the 1870s due to their geographic location. They were at the northern limits of Spanish colonization. The Franciscan attempt to establish missions during the seventeenth century failed when the pueblo uprising of 1680 expelled them. As a result, the ethnoastronomy studies of the Hopi in the 1880's and 1890s probably have the greatest longevity and strongest connections to prehistory cultures such as the Sinagua.

The concept of time to the ancestral Pueblo cultures was probably more of an organic experience than a mechanical one. They felt the passage of time differently than we do today. Without clocks and wristwatches to mark the hour, the ancestral Pueblos lived time as each day steadily changed in length throughout the year. They would have relied on their "body clock" to determine their daily activities. Time was not measured by a daily calendar but rather by the events that occupied it. In an exhaustive study of

9

Hopi linguistics and the concept of time, Ekkehart Malotki (1983: 632) provides an ethnographic analogy to the ancestral Pueblo when he observed that:

> "... the Hopi Indians lack neither an elaborate consciousness of time nor its reflection in their speech . . . we can also say that their sense of time, or the role that time plays in their lives and culture, does not correspond to ours. Nor would one expect the two to be identical. . . . Time-reckoning methods, calendrical systems . . . are very complex and highly sophisticated in both the Hopi and our western world."

Anthropological studies have long recognized the Hopi calendar as being based upon careful observations of the sun and moon (McCluskey 1977). The primary goal of these observations was to develop a calendar that would serve to regulate agricultural activities as well as the religious ceremonies that defined their community. Their astronomical system was based on observing the phases of the moon or the rising and setting sun against known landmarks on the horizon.

The observations were often the responsibility of a religious leader called the **Sun Watcher** (Zeilik 1985). Most Hopi villages have their own sun watchers. This independence in sun watching is in line with the independence of each Hopi village for ceremonies. The sun watchers are traditionally of the Water Clan (McCluskey 1990). By their readings of the positions of the sun, they would know when the various ceremonies and plantings were to take place.

> " . . . my father, Sitaiema, was a Water Clan man . . . he was the official Sun Watcher for the village. . .. His observatory was out at the point of the mesa, at the south end. He had a notched stick, with the notches in groups of four, and he'd hold this notched stick in such a way as to measure the horizon and see exactly where the sun was coming up. . . . His sun readings set the calendar for all important events and indicated to the different clans when they were supposed to do various things. . . . He announced everything through the year, and the whole cycle of ceremonial events was set by his sun readings." (Yava 1978: 72-73)

All Hopi ceremonies follow a similar pattern (McCluskey 1977). On the evening that the sun reached a specific position, or the Moon has entered a specific phase, the sun chief will call a meeting during which the leading members decide the date of the ceremony. The date is announced at sunrise the following morning. Zeilik (1985) points out that the "anticipatory aspect of sun watching is the most important feature of the sun watcher's job." This is because material and spiritual preparations are required for each ceremony.

A CAUTION!

Although this study will suggest ethnographic analogies for the calendric events at the V Bar V site to prehistoric and historic Pueblo agricultural and ceremonial activities, it is done with caution. As noted archaeoastronomer Michael Zeilik (1985) pointed out: "It is difficult and dangerous to generalize about Pueblo society. Each village tends to act pretty much on its own." Villages differ in the timing and importance of some ceremonies and even differ as to whether they are solar or lunar based. Studies even suggest that the number of months in the year varied from seven, to eight, to twelve, to thirteen, depending upon the village, while some villages even had "nameless months" (Spier 1955).

It must also be accepted that historic Pueblo societies have been participants in the evolving traditions of their Southwestern cultures from the time of the prehistoric Pueblos. It would be unsupportable to contend that innovations and modifications were not made by historic cultures as the rituals and practices were passed along. Interaction among villages, different cultures, as well as the rise of new leaders and ceremonial concepts, would be expected to bring change and the diffusion of the original intent.

Because of the passage of time and the evolutional tendencies of traditions, it was not expected that close similarities would be found. However, some basic similarities between prehistoric and historic cultures, as evidenced in rock art, pottery and agricultural practices, can be drawn.

This study focused on Hopi ceremonies since both Hopi and archaeologists recognize Sinagua as being one of the groups that is ancestral to Hopi. Since astronomical associations have been documented at Hopi as indicators for the scheduling of their ceremonies, similar associations were likely present in Sinagua society. Even though many of the Hopi ceremonies are directly related to the Katsina religion, a world-view religion that does not come into existence until after the demise of the Sinagua as a viable cultural group, it is likely that similar observations and ceremonies took place before the advent of the Katsina religion. While there is no known direct link between Sinagua beliefs and the Katsina, it is plausible that given the Hopi ancestral ties, something likely preceded the Katsina ceremonies that provided a foundation for the Katsina religion to build upon (probably the various religious societies that were certainly in existence by the time the Katsina religion comes around).

Some believe that rock art, of any type, should never be interpreted since no direct descendants remain and only someone of that culture could understand their intent. However, Whitley (2005) has suggested that the "archaeological record is in fact a record of human behavior and there is no reason why meaning cannot be derived from archaeological analyses."

This survey will show that great care was taken to mark the passage of time and the arrival of specific points in time. This will tend to support the primary hypothesis that a cycle of annual, calendrical-based rituals and practices were associated with the Sinagua of the Verde Valley.

3

ARCHAEOASTRONOMY BACKGROUND

ASTRONOMICAL BACKGROUND

For a time many believed that the Earth getting slightly closer to the sun, resulting in higher temperatures, caused the seasons. We now know that the real cause of seasons is the angle of the Earth's rotational axis (23.5°) to its orbital plane as shown in Figure 4. The hemisphere of the Earth that is tilted toward the sun receives a greater amount of solar energy than the hemisphere tilted away, resulting in higher temperatures. In the northern hemisphere, the summer solstice occurs when the sun is farthest north, while the winter solstice occurs when the sun is farthest south. An equinox occurs when the sun is half way between the solstices.

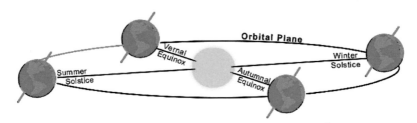

Figure 4. Seasonal changes in the northern hemisphere

The Earth's orbit around the sun is not circular but elliptical. This causes the distance from the sun to vary annually. The Earth is closest to the sun (perihelion) on January 3 and farthest (aphelion) on July 4.

The annual change in the relative position of the Earth's axis in relationship to the sun causes the height of the sun (**altitude**) to vary in the sky. When the sun is on the horizon at sunset or sunrise it has an altitude of 0°. When the sun is at the highest point in the sky (its **zenith**) it is referred to as ***solar noon***. Altitude is usually used together with azimuth to give the direction of the sun. ***Azimuth*** is the direction of the sun measured clockwise around the observer's horizon from north. The Sun has an azimuth of 0° due north, 90° due east, 180° due south and 270° due west. These terms are represented in Figure 5.

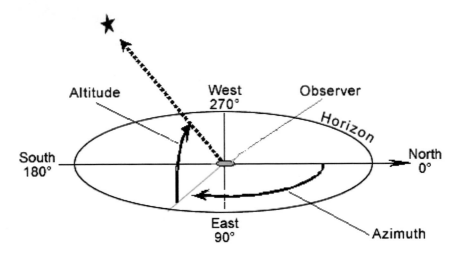

Figure 5. Solar altitude and azimuth

The V Bar V observations employ the midday sun. The solar altitude thus becomes a useful measure for predicting events. Figure 6 provides the altitude variations for the equinox and solstice presentations at the V Bar V solar panel.

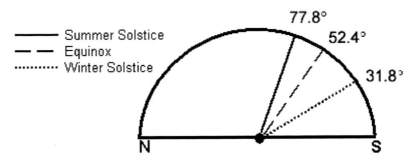

Figure 6. V Bar V variations in solar altitude at
the solar panel during solstice and equinox

Measurements of azimuth are also useful for the V Bar V observations. The astronomical effects at V Bar V last an average of six minutes, the time it takes for the azimuth of the sun to change by 4° to 6° (Figure 7). The effects are then diluted by the sun's changing declination.

By calculating the altitude and azimuth at the V Bar V latitudinal position, we can calculate (and predict) the onset of the solar events. Figure 8 provides the times of solar noon and onset of the sun shaft as observed or predicted from the vernal equinox through the autumnal equinox.

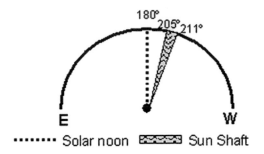

Figure 7. V Bar V solar azimuth at solar noon
and range during sun shaft appearance

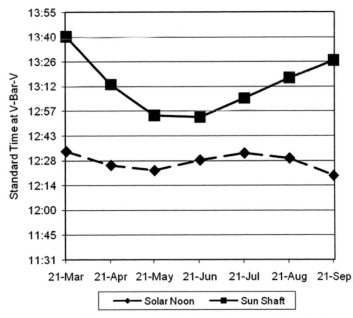

Figure 8. Solar noon and mid-point and onset of sun shaft

The complete calculation of altitude and azimuth at the V Bar V between equinox events is provided in the Appendix A.

EQUINOX AND SOLSTICE OBSERVATION DATES

This study monitored the site over a full 12-month cycle of the sun. The primary observations were around those events readily apparent through naked-eye observations, namely the equinoxes and solstices. But the date of these events fluctuates. For example, the vernal equinox occurs on dates varying from March 19 to March 21.

As described in Chapter 2, it is known that the Sinagua lived in the Beaver Creek area between A.D. 800 and 1400. It is not know exactly when the astronomical elements were created. From the degree of repatination of the elements on the panel it appears that they were not the earliest, nor were they the latest. As a median, the time period between A.D. 1000 and 1200 was used to calculate the dates and times of the equinoxes and solstices as shown in Appendix B. The 21st of each month was selected as the general mean date for observations, although the 2005 astronomically determined dates were observed as well.

While conducting this yearlong survey, some visitors to the site have asked -- "isn't the sun in a different position today than it was when the images were made?" The short answer is "yes." Celestial objects change the positions of their rise and set slowly with time. This is referred to as the ***obliquity of the ecliptic***. Since the days of the Sinagua at the V Bar V site, the change in the obliquity has shifted the position at which the sun rises and sets by only 0.07 degrees.* This would not produce a significant change in the observed sun/shadow patterns of our midday observations. If we were attempting to replicate the markings of the sunrise position on the horizon, it would be slightly different, and this change would need to be taken into account.

PREHISTORIC SKYWATCHING TECHNIQUES

Archaeoastronomy studies involve naked-eye observations that relate astronomically significant dates with the orientation of an ancient place, setting, structure, pictograph, petroglyph or rock. Because the variations in solar altitude cause the seasons, the azimuth variation can be used to mark them. A good case can be made that this was done by Paleolithic and Neolithic cultures (Kelley and Milone 2005).

To determine what these cultures could have learned just from their observations of the heavens over many years, without the benefit of today's

---

* Calculation courtesy of Lowell Observatory, Flagstaff, Arizona

astronomical knowledge, we must look through their eyes. The Sinagua, for example, could have known with naked-eye observations when summer begins (the sun is the highest in the sky for the year, and the day is the longest) as well as when winter arrives (the sun is the lowest in the sky for the year, and the day is the shortest). They could also have determined the equinoxes (the time when day and night are about equal).

Sinagua sunwatchers, similar to historic Pueblo sunwatchers, most likely monitored the sun and anticipated the times of equinoxes and solstices through *imaging* and/or *sighting*. In the vast majority of situations, they would have recorded these events by sighting sunrise or sunset locations on the horizon (Figure 9). For such horizon studies, the azimuth (the direction of the sun at the observer's horizon measured in degrees from true north) is the astronomical measure for those events. The sun would appear on the horizon at various locations as the seasons progress.

Figure 9. Horizon sighting method of calendrical observation

The predominant method of watching the sun at the historic Pueblos was to use horizon markers from a fixed observation position. Horizon calendars at Hopi have been documented for each of the three mesas (Forde 1931; Stephen 1936; McCluskey 1977; Zeilik 1985a).

In addition to sighting, the observers took notice at these important times of when sunlight entered their rooms or illuminated specific images. This imaging technique of calendrical observation (Figure 10) involved the sun casting its effect on an image or architectural feature at sunset, sunrise or during the day.

Figure 10. Sunlight illuminates concentric circles at V Bar V

Solar calendars based on this imaging method have been identified at Ancestral Puebloan sites including Chaco Canyon, Hovenweep, Chimney

Rock, Yellow Jacket and Mesa Verde (Williamson 1984). Similar solar markers have been identified at Northern Sinagua locations at Wupatki (Bates 1996). Preston and Preston (1987) identified eighteen potential Ancestral Puebloan solar marker sites in or near the Petrified Forest National Park. Solstice and equinox markers and shrines have been found in the Mimbres/Mogollon region (Ellis and Hammack 1968; Kriss 1989). In addition, Bostwick (2002) recorded Hohokam calendrical sites at twenty-two locations in the Phoenix region that employed architectural alignments, imaging locations and horizon markers. The Southern Sinagua were virtually surrounded by cultures using solar calendaring techniques.

While there have been anecdotal reports of solar calendrical petroglyph/pictographs within the Southern Sinagua areas of the Verde Valley none have been fully documented. This study documents what is believed to be a 12-month solar calendar of the Southern Sinagua using the midday sun.

SHADOW STONES

In addition to sunlight effects, there are also shadow interactions at the V Bar V site. Shadow and sun shaft effects upon the V Bar V Solar Panel are created by a trilogy of boulders wedged in a crevice in the rock face that separates the solar panel from the petroglyph panel to its south. The boulders act as *gnomons* which are objects, such as a sundial, that projects a shadow used as an indicator of time. .As shown in Figure 11, two boulders protrude from the cliff face while a third acts as a wedge to hold them in place.

Figure 11. Gnomons or shadow-casting stones

Because it was impossible to accurately assess the nature of the rock gnomons when viewed from ground level, an archaeological research permit was approved by the U.S. Forest Service to erect a temporary scaffold permitted an examination of the critical rock face up close (Bostwick, Lindberg and Zoll 2011) . The scaffold was erected so as to not touch or damage the rock face.

As the result of this close examination, it was determined that all of the gnomons are considered to be naturally occurring boulders that have been frost heaved outward to their observed locations. Each block came from within the rock fracture and were not placed there by human hands. A surprising finding was that humans had purposely placed small rock wedges (Figure 12) into the bounding edges of all three of the critical rock blocks in an effort to stabilize their further movement.

It was also noted that a portion of the surface of right gnomon appears to have been removed through flaking and/or prying for the purpose of controlling the shape of the gnomon's shadow. Sandstone is difficult to tell if it has been flaked because its rough texture does not always produce clear flake scars, especially after it has been exposed to the elements. None-the-less, portions of heavily patinated areas on this boulder have been removed along its upper edges. Faint flake scars appear to be visible, as well as a couple of indirect percussion scars where the blow was struck by a punch to remove flakes.

Figure 12. Wedge stones
(Photo: Todd Bostwick)

ASTRONOMICAL PETROGLYPHS

The V Bar V Heritage Site contains about 1,035 petroglyph images. Continued study of the site and recent excavations will most likely result in additional image recordings. It is the largest petroglyph site in the Verde Valley. The archaeological features of the site are attributed to the Southern Sinagua culture that occupied the area.

The site contains a number of images that are generally ascribed as having astronomical meaning (Figure 13). These include spirals, concentric circles and sun-like glyphs (Patterson 1992). Snake-like images, though generally not associated with astronomical meaning, had a close association with the sun (Serpent of the Heavens) in historic Pueblo cultures and were also connected with fertility and equinox (Tyler 1964).

The panel under study contains 10 of these spiral, concentric circle and snake-like glyphs. There is also a very unique image that incorporates a sun-like glyph with a pair of arched lines. Table 2 provides the dimensions of each symbol.

The overall width of the designated imagery (image #1 to #7) is 2.9 meters (9.5 feet). The expanse from the highest image point (#8) to current ground level directly below is also 2.9 meters although this measurement does not accurately describe the height at the time the images were created. Rock falls from the cliff overhead and flooding by neighboring Wet Beaver Creek over the past 600 years have deposited silt and rock at the cliff face, estimated to measure .8 to 1 meters (2.6 to 3.2 feet) from the time the images were created.

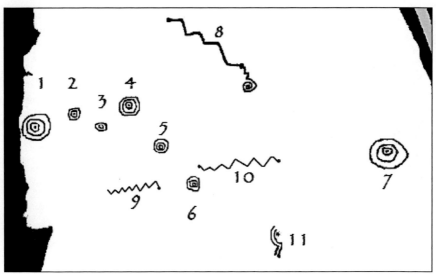

Figure 13. V Bar V astronomical symbols

The most prevalent astronomical glyph is that of the concentric circles. These are composed of two or three circles with a central depression.

> "In Pueblo explanation of this old symbol, so standardized as possibly to warrant designation as a glyph, the outer circle represents the ring of light around the Sun, the second represents Sun himself, and the inner circle or dot his umbilicus, which opens to provide mankind with game and other sustenance." (Ellis and Hammack 1968: 35)

Table 2. Astronomical symbols
(width x height)

Concentric Circles	Falling Spiral
1 – 23cm x 21cm	8 – 8.5cm x 7cm x 83cm
2 – 9cm x 8.5cm	Snake-like glyphs
3 – 8cm x 8cm	9 – 47cm
4 – 14.5cm x 14.5cm	10 – 64 cm
5 – 8.5cm x 8.5cm	Sun-like glyph
6 – 10cm x 10cm	11 – 8cm x 24cm
7 – 26cm x 21.5cm	

Another possible example of this symbolism in Sinagua iconography may be represented by one of the artifacts found during an excavation in the 1930's at a 12th century pueblo near Flagstaff. At this site was found the burial of a man whom Hopi identified as an important person within several religious societies. Among the items found with him was a disc showing three concentric circles of white shell inlay which enclosed a central white dot – again, the Pueblo sun symbol (McGregor 1943).

While these examples show the significance of this symbolism to the prehistoric and historic puebloans, why was it chosen to represent the sun? The explanation is actually quite logical to anyone who spends their days observing the sun. The figure below illustrates how the Sunwatchers would observe the eastern sunrises and western sunsets of the solstices and equinox. They would have noticed that the sun moves in three arcs across the sky during the day.

The Hopi and other pueblos believe that the sun has two houses; one in the east from which the sun emerges at sunrise into the Upper World and one in the west where it descends at sunset into the Lower World (Young 2005). This belief is part of an organizing principle of duality in the pueblo world. The concentric circle image, therefore, represents the path of the sun through the upper and lower world, adding to the duality concept of two equinoxes and two solstices.

Snakes are important in Hopi during the Snake Dance ceremony, and the great Water Serpent is a deity to both Hopi and Zuni. The serpent appears with sun symbols (concentric circles) at the Hovenweep National Monument in the Holly House solar calendar (Williamson and Young 1979). Further, a snake-like image has been documented with a concentric circle as part of the sunrise equinox markings at Fajada Butte in Chaco Canyon, New Mexico (Sofaer et al. 1983).

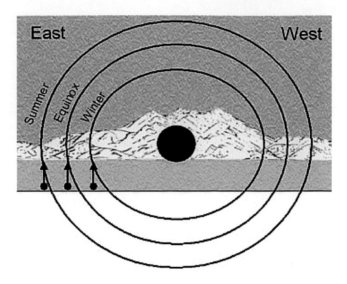

Figure 14. Potential concentric circle solar origin

SACRED MOUNTAIN SUN SHRINES

One of the questions that arose during the survey of the petroglyph panels is how did the Sinagua know what day to record the petroglyph images so that they would coincide with the equinox and solstices. The most obvious answer was that they observed the horizon and transferred that knowledge to the petroglyph panel. To confirm this hypothesis, the ruins atop Sacred Mountain were visited at sunrise on the equinoxes and solstices. The positions of the rising sun along the Mogollon Rim were recorded on those days. Over the next several months, those positions were visited. At each of these locations, sun shrines of the same construction were round at the equinox and solstices positions.

As pointed out by Zeilik (1985b: S91) "one of the important aspects of Hopi Sun shrines is their location on the tops of mesas at places that mark important times of the solar calendar when viewed from the Sun-watching station of the Sun Priest." These shrines may consist of a natural or man-made pile of rocks. The shrines that we discovered consisted of a pair of stone piles. When standing between these piles the declination was an exact straight line from the sun-watching station to the position of the rising sun. These twin piles gave the appearance of a gateway, or portals, through which the sun would rise on these important calendric dates. The discovery of these shrines at the expected locations confirmed that prior to the creation of the petroglyph calendar, the Sinagua sunwatchers used horizon sighting for their calendric purposes.

4

OTHER FEATURES OF INTEREST

CORN PLANTS

A particular petroglyph element that drew attention during the survey was a series of images that have been referred to as "centipede-like" (Figure 15). This image, however, is very similar to others that have been interpreted as a maize or corn stalk. Mallery (1893) noted that "The object resembling a centipede . . . is a common form (of corn symbol) in various localities of Santa Barbara County, California . . . . In other parts of Arizona and New Mexico petroglyphs of similar outlines are sometimes engraved to signify the maize stalk." Similarly, Colton (1946), in describing Picture Rock near Willow Springs, Arizona, reported that a "well-informed Hopi Indian" had identified a similar image as a "corn plant." This site is known at Hopi as "Tutuventiwngwu."

Figure 15. Corn plant highlighted images

Peter J. Pilles, Jr., Forest Archaeologist for the Coconino National Forest, observed that what is interesting here is that the "centipede" form is often said to be an archaic form. A good example can be seen in the classic dark purple archaic paint at the Grotto Site at the Red Cliffs/Palatki Heritage Site, northwest of Sedona. But he noted that here we see the element in an indisputable later context. Is this a long-term retention of an early element, whatever it depicts, or an indication that the element is a late archaic depiction of corn that appears and becomes entrenched as an icon at the time that corn is introduced into the Southwest from Mexico?

The repeated depiction of such a long-lasting and widespread image caused a closer examination of whether this non-astronomical symbol on the Solar Panel would hold any relevance.

WATER CLAN IMAGES

One of the generally recognized functions of rock art is clan designation (Weaver 1984). The V Bar V petroglyph panels have several images associated with such clan designation. The solar panel in our study has quadraped images centered near the top of the solar panel (Figure 16). These have been ascribed as a possible Water Clan symbol. As noted earlier, the Water Clan provided the official sun watchers. The placement of two such clan symbols so prominently on the solar panel adds another suggestion of the astronomical nature of the panel.

Figure 16. Possible water clan symbol

As the study of other potential Sinagua calendar sites in the Verde Valley continues, a pattern is emerging. The same image is present in pictograph or petroglyph form at several of these sites. It has also been found at the Northern Sinagua calendar sites of Wupatki (Bates 2005) and Chavez Pass (Snow 2006). This has become one of the "clues" to watch for as potential calendar sites are investigated.

Today, the Hopi are divided into several ***phratries*** (a set of related clans). The Patki (Water-house or Cloud) phratry has been associated with the horned-toad image. The Patki Clan is said to have arrived at Hopi from the Pala'tkwabi Trail (Byrkit 1988) which passes through the Verde Valley. Further study and analysis of this clan's origin and potential linkage is being pursued.

SLIVER STONE

There is an interesting sliver of sandstone (see Figure 1 on page 4) wedged between the Solar Panel and the panel to its south. The edge of the sandstone appeared to have been "worked" to produce a jagged edge.

The construction of the 2011 scaffolding, referred to previously, was instrumental in determining whether this edge had, in fact, been manufactured. Geologist Paul Lindberg concluded (Bostwick, Lindberg and Zoll 2011) that the rock sliver had definitely been modified by man. First of all, the acute leading edge of the rock sliver has been modified by hammer blows directed from the right (south) side. The result is similar to that formed while flint knapping and removing flakes, but in this case it is sandstone that is being chipped off. Secondly, the right side (south) of the rock sliver has been obviously smoothed with near vertical scratch marks. The uppermost part of the rock sliver has been smoothed and the former natural pitting of the sandstone, with its surface manganese patina, had been removed above but preserved below. The effect is to "sharpen" the acute edge of the rock and smooth the surface out.

Furthermore, Todd Bostwick (Bostwick, Lindberg and Zoll 2011) concluded that not only was it flaked, but a combination of direct percussion and pressure flaking techniques were used. Because the thin edge is very fragile, it seems likely that soft-hammer percussion, utilizing an antler billet or a wooden baton, was used to remove the larger flakes rather than a stone hammerstone which can be quite destruction on more fragile materials. Soft-hammer percussion usually affords the knapper (lithic tool maker) more control over the size of the flakes that are removed because the hammer is more elastic and produces flakes that are larger and thinner with small bulbs of percussion. However, a skilled and experienced knapper can use a small stone hammerstone with equal results

As the cliff face is approached, this light red sliver of stone (Figure 17) stands out starkly against the surrounding dark desert varnished bluff. This was apparently the effect that was intended.

Figure 17. Sliver stone image presented horizontally

Such "edge workings" have been noted at other petroglyph sites both in the Verde Valley (for example, in Rarick Canyon) and elsewhere in Northern Sinagua areas (at several sites in Anderson Pass and also at Chavez Pass). Exactly what these edge markings are all about has never been satisfactorily explained.

The sun highlights the edge workings at V Bar V in the late afternoon, producing a mirror image in shadow form (see Figure 16 again). When viewed on a 90-degree angle, the shadow can appear to present a mountain range horizon, which we know were important as markers for celestial events. Further study is required of these edge workings to determine any similarity in construction and meaning.

Figure 18. Close-up of worked edge
(Photo: Paul Lindberg)

5

VERNAL EQUINOX SEQUENCE

OVERVIEW OF OBSERVATIONS

While studying the petroglyphs at the V Bar V site, the quantity of concentric circle images on a single panel was noted with interest. Knowing how similar images were used at other sites in the Southwest to identify solar calendars, I began to study how the sun might interact with these images. The appearance of the sun shaft and shadow lines, described in the previous chapter, was first observed in February 2005. The slow progression of the sun shaft down the panel toward the petroglyphs suggested that some interplay with the petroglyphs might be about to occur.

The effect of the sun on the petroglyphs was first recorded at the vernal equinox of 2005. Observations were followed up on the 21st of each month thereafter. This was done as a rock art experiment to track the sun pattern over the period of a full year and to see what light and shadow interactions occurred during the year in order to determine if they were coincidences or an intentional pattern. During the course of the observations other dates of potential ceremonial significance were also observed.

During each observation, from the vernal equinox to the autumnal equinox, it was noted that at least three and as many as six petroglyphs were involved in either the sun shaft or shadow lines. Table 3 provides a summary of the petroglyphs that were found to have potential interactions with the sun shaft or shadow lines. Figure 19 shows the Solar Panel with the images that were tracked enhanced to show their relative position. As will be shown, no petroglyphs are involved in observations from October through February.

Table 3. Summary of petroglyphs involved in astronomical events

Image	Equinox	April 21	May 21	Summer Solstice	July 21	August 21	Equinox	Winter Solstice
1	X	X				X	X	
2	X						X	
3		X	X		X	X		
4	X	X	X	X	X	X	X	
5								
6		X	X	X	X	X		
7								X
8		X	X	X	X	X		
9			X	X	X			
10		X	X	X	X	X		
11				X				

Figure 19. Solar Panel with tracked images enhanced and numbered

MARCH 21 -- VERNAL EQUINOX

The time of the vernal equinox is still an important event for today's Hopi. We can assume that it was also important to their ancestors, the Sinagua. When the sun illuminates the sun symbol, the Sunwatcher of the Water Clan would know that it was time for the vernal equinox activities. Today, in some villages, these take the form of night dances to create a pleasant atmosphere for all life and to encourage growth, and to bring all-important rain for fruitfulness.

Figure 20. Solar Panel illuminated by the sun shaft on March 21

As mentioned earlier, the astronomically determined date of the vernal equinox can vary from year to year between March 19 and 21 (Appendix B). The actual date of the equinox in 2005 was March 20 at 5:33 a.m. local time, however the site was cloud covered, preventing any observation. Observations were successful on March 21. The sun's azimuth position for the two days were identical at the time of observation but presented one minute later on March 21 than would have been presented on March 20. The altitude difference between the two days at the point the sun crested the bluff was 0.5°.

The two shadow stone outcroppings initially cast shadows over the entire panel except for a shaft of light between them. Gradually, two distinct shadow lines were produced (Figure 20). The sun shaft highlighted a major concentric circle image (#1). In addition, a secondary concentric circle image (#2) was tangent to the top of the shadow line (Figure 20) while another major circle image (#4) was tangent to the bottom of the shadow line.

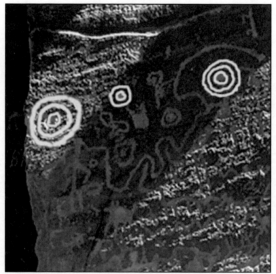

Figure 21. March 21 observation enhanced

The effect began at 1:41 p.m. and lasted for six to seven minutes before beginning to brighten to full sunlight as the sun moves over the bluff. As Williamson noted (1984: 94) during his observations of the summer solstice marker at Holly House "it takes only seven minutes for this compelling drama of light and shadow to play out, but in that time even the modern secular observer is drawn into it, feeling a sense of wonder and an unusual oneness with the cosmos."

APRIL 21

At first, the intention of the observation on this date was to simply record the passage of the sun shaft until it approached the summer solstice. As expected, the effect presented itself in a lower counterclockwise position due to the rising position of the sun. The effect began at 1:11 p.m. and again lasted for six to seven minutes.

First, it was noted that the sun shaft (as seen in Figure 22) was the exact width of the space between two "notches" in the outer edge of the bluff face. Several notches in the outer edge provide interesting effects during many of the observations. If you look back to Figure 20, you will notice that the bottom of the lower shadow line is in the exact same position as the bottom of the top shadow line in Figure 22. Close examination of these chips suggests that they are natural.

Figure 22. April 21 observation

As observations continued, it became apparent that significantly more was occurring than had been anticipated. A total of six images were either in the sun shaft, bisected or in tangent positions (Figure 23). The large concentric circle (#1), that was in the sun shaft for the vernal equinox, was now bisected. In addition, the next large concentric circle (#5) was now in the sun shaft, while a small concentric circle (#4) was tangent to the top stone's bottom shadow edge.

Further, two images (#8 and #10) were clearly touched by sun or shadow lines. The sun-shaft begins at the point of a circular depression in the falling spiral image (#8) and extends exactly over three sections of the image. The spiral at the end of the image was also tangent to the base stone's bottom shadow line. This same shadow line touches the circular depression at the beginning of the snake-like image (#10). This shadow line continues to bisect a small concentric circle (#9).

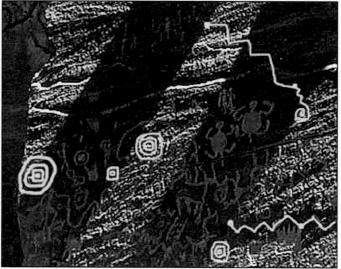

Figure 23. April 21 observation enhanced

Successful farming, especially in the drier climates, depends on knowing when to plant various crops. April 21 marks the beginning of the third week of April. This is the planting time for the early corn crop of the historic Pueblos (Forde 1931). Forde noted that this was of great importance since it provides corn for the Niman ceremony after the summer solstice.

The elaborate interplay of these six images suggests that great care was taken to ensure that the proper early planting time was calculated correctly.

The care taken to ensure that the sun shaft highlighted three segments of the falling spiral also appears to be significant and is discussed more fully in Chapter 9.

The corn planting connection was further strengthened by the inclusion of one of the corn plants in the bottom shadow line of the base stone, as seen in Figure 24. The shadow line touches the end of the corn plant, which is in perfect alignment with the head of the snake-like image (#10) and the bisection of the concentric circle image (#5). This would be consistent with the beginning of the corn planting season.

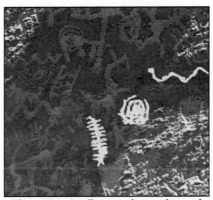

Figure 24. April corn plant enhanced

MAY 21

The event on May 21 continued on the counterclockwise path (Figure 25). The effect began at 12:55 p.m. and lasted for six minutes. It again presented a complex interplay of images as had occurred in April. A total of six images were either in the sun shaft, bisected or in tangent positions (Figure 26).

Two concentric circles (#3 and #4) were tangent to the upper edge of the shadow line produced by the top stone, either in the sun or in shadow. The falling-spiral image (#8) was cut by the lower edge of the top stone shadow line precisely at the joint of the fourth and fifth segments. The fifth and sixth segments of this image were perfectly framed by the sun shaft. You may recall that the April 21 image highlighted the first three segments.

Figure 25. May 21 observation

One of the most interesting effects was that the width of the top-stone shadow was the exact width of a snake-like image (#9).

The base stone's top shadow line bisected a concentric circle (#6). It was also tangent to the circular depression in a snake-like image (#10), while the lower edge of the shadow line bisected the ninth and tenth segments of the image.

Again, it was interesting to note that the upper edge of the top stone's shadow line (as seen in Figure 25) was on the lower of the two notches in the outer edge of the bluff face that was noted for April 21. The south edge extended to current ground level. Whether the bluff edge contains a feature below ground that may have been highlighted will have to await future excavations.

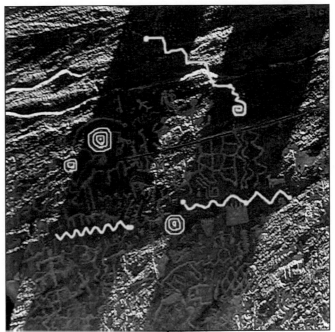

Figure 26. May 21 observation enhanced

Once again, the strength of the corn planting connection is made by the inclusion of a corn plant image within this formation. As seen in Figure 27, one corn plant is wholly within the sun shaft.

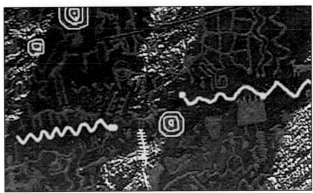

Figure 27. May corn plant enhanced

As the weather warms in the third week of May (beginning May 21), it is time for the main planting of corn, beans and squashes (Forde 1931). Once again, we see a very elaborate construct of images that can only be attributed to a need to be precise with the message it presents. The use of the snake-like image (#9) to mark the width of the top stone's shadow with such precision attests to the deliberate intent of ensuring the proper reading. Since this is the time for the primary crop planting, it would have been essential to accurately determine the most beneficial time for this activity. This would seem to be suggested by the appearance of the corn plant in the center of the sun shaft.

6

SUMMER SOLSTICE SEQUENCE

JUNE 21 -- SUMMER SOLSTICE

The Summer Solstice usually occurs on June 21 or June 22 (Appendix B). The solstice in 2005 was on June 20 at 11:46 p.m. Observations were made on June 21. The Sun continued on the counterclockwise path (Figure 28). The effect began at 12:51 p.m. and again lasted for six minutes before the base-stone shadow began to recede. It again presented a complex interplay of images as had occurred on the previous observations. A total of six tracked images were either in the sun shaft, bisected or in tangent positions (Figure 29).

Two concentric circles (#4 and #6) were tangent to the upper and lower edges of the shadow line produced by the top stone. They appeared to frame the shadow. The falling spiral image (#8) was cut by the lower edge of the

Figure 28. June 21 observation

top stone shadow line precisely at the joint of the fifth and sixth segments. Further, this image was cut by the upper edge of the base stone's shadow edge precisely at the circular depression at the beginning of the rapid fall to the spiral. The two snake-like images (#9 and #10) continued to be dissected by the shadow lines at various segment joints.

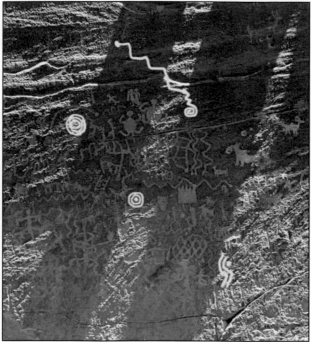
Figure 29. June 21 observation enhanced

Once again a corn plant is touched (Figure 30), this time in tangent to the top of the sun shaft. This may have indicated the end of the corn planting season.

Figure 30. June 21 corn plants enhanced

One of the most interesting effects was that the bottom edge of the base stone's shadow was tangent to the sun-like glyph (#11). Note that a secondary narrow shaft of light is becoming more distinct and is approaching this image (Figure 29).

With the summer solstice, the sequence of plantings ends. For the historic Pueblos, this signifies a major turning point in the year from a season of preparation and planting to a season of fruition and harvesting (McCluskey 1977). The change is commemorated by the Niman, or Home Dance, ceremony. The date of this ceremony is announced on the Summer Solstice. All of the Katsinas have been with the people on the earth since February to help establish the growing season for the year and it is time for them to return home to the San Francisco Peaks. The ceremony culminates in late July with the Home Dance. They dance and sing to symbolize the harmony of good thought and deed – a harmony that is required for rain to fall and for a balanced life. With the performance of the Home Dance the Katsina spirits depart the earth and return to the spirit world.

As mentioned earlier, there is no archaeological evidence that the Sinagua had the concept of Katsina. The earliest record of the Katsina religion occurs in the 14th century. However, it is feasible that as the Sinagua migrated to the Hopi mesas, they brought with them the ceremonial practices and beliefs that may have contributed, in part, to the eventual Katsina religion. As Hays-Gilpin (2006: 21) has described, Hopi will sometimes refer to early cultural artifacts "to a time when the people were still 'becoming Hopi.' In their own view, Hopi history is not a history of one people but a series of histories of clans who came together to become Hopi people – Hopisinom. In a philosophical sense, 'Hopi' is not a tribe, an ethnic group, or a language, but a way of life."

The announcement of the date for the Niman ceremonies on the summer solstice brings a potential interpretation to the unusual sun-like glyph with the arched lines that only comes into the astronomical cycle at this time. The arched lines are reminiscent of dancing figures. The fact that a shadow line on the summer solstice first touches this image suggests that it was intended to mark the beginning of preparations for the ceremonies. To determine the feasibility of this interpretation, the image was observed for several days.

JULY 8

As mentioned above, the date of the Niman ceremonies are announced on the summer solstice. This announcement is followed by sixteen days of prayer and meditation. To determine whether the solar panel contains ceremonial markings to indicate the beginning of this festival, the site was visited every few days.

On July 8 the noted images were highlighted at 12:57 p.m. as shown in Figure 31. The effect again lasted 6 minutes. Attention was focused on the suggested Home Dance glyph. As shown in Figure 32, the image was perfectly outlined by the second sun shaft. It is conjectured that this may have indicated the day on which the Niman ceremonies were to begin. There are 16 days (the period of prayer and medication) between June 21 and July 8. The glyph was observed on July 10 as well and it was documented that the sun shaft was no longer perfectly outlining the Home Dance glyph.

Figure 31. July 8 observation

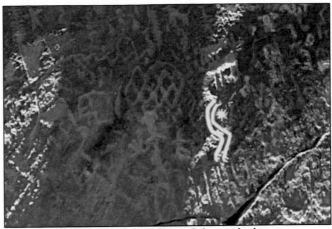
Figure 32. July 8 enhanced dance glyph

JULY 21

After the summer solstice, the earth's orbit makes the sun appear to reverse direction and begin a descent in the sky. Following the contention that the Niman ceremonies were to begin on July 8, we began to observe for indications as to when the ceremonies might conclude. The day following the conclusion of the ceremonies is the traditional date for the first sweet corn harvest.

The event recorded on July 21 at V Bar V was the first observed on the sun's reverse course, now on a clockwise path (Figure 33). The effect began at 1:04 p.m. and lasted for six minutes.

A total of six images were either in the sun shaft, bisected or in tangent positions. As expected, it presented the same complex interplay of images as had occurred on May 21 (Figure 25), with one exception. At that time, the secondary sun shaft to the south of the main sun shaft did not extend as far down the panel. However, on July 21, the second sun shaft now framed the southern-most corn plant (Figure 34).

McCluskey (1977) observed that the Niman festival culminates with a public dance (the "Home Dance") on the ninth day. He assembled a range of dates from various studies that reported the "ninth day" as occurring on July 27 in 1891, July 21 in 1892, July 23 in 1893 and July 27 in 1921. Of course none of these dates occur nine days after July 8, but the range of dates reported for these four events display an uneven application.

Figure 33. July 21 observation

From the previous studies that sought to record the Niman start and ending dates, the obvious conclusion is that there was confusion and contradiction among the researchers at the time of those studies. It is unlikely that we will ever be able to conclusively determine the dates used by the Sinagua if they observed this or some similar ceremony that became the Niman. Therefore any attempt to link the solar effects at the V Bar V with this festival will have to remain conjecture. Whether the Sinagua performed a predecessor ceremony to the Niman ceremony for nine days, for some other number of days, or not at all, is unknown. There are, however, some tantalizing similarities in ceremonial timing and historical custom.

What can be stated with assurance, however, is that on July 21 two sun shafts framed two corn plants. This is a little over 90 days from the first planting of corn that began on April 21. The average time required for corn to mature in the Verde Valley is 63-100 days, depending on rainfall totals*. The July 21 markings, therefore, may have indicated the time for the first corn harvest. Forde (1931) indicated this was of great importance as

*Source: University of Arizona Extension Service

42

part of the Niman ceremony. Similarly, McCluskey (1977) pointed out that the first sweet corn of the season was harvested the day following the return of the Katsinas to their home.

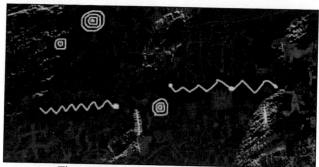

Figure 34. July 21 corn plants enhanced

AUGUST 21

The event on August 21 began at 1:16 p.m. and lasted for six minutes (Figure 35). The sun continued on its clockwise decent to the south. As expected, it presented the same complex interplay of images as had occurred on April 21 (Figure 22). Six images were either in the sun shaft, bisected or in tangent positions (Figure 36). The outer edge notch was similarly the width of the sun shaft.

Figure 35. August 21 observation

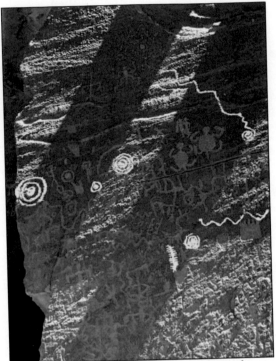

Figure 36. August 21 observation enhanced

With the touching of the corn plant by the sun shaft, the corn from the May 21 planting has reached the middle of the maturity range. This may have indicated that harvesting of the main corn planting could begin if sufficient rains had arrived to bring the plants to maturity.

7

AUTUMNAL EQUINOX SEQUENCE

SEPTEMBER 21 – AUTUMNAL EQUINOX

The autumnal equinox can occur at any time from the 22nd to the 24th of the month (Appendix B). The Autumnal Equinox in 2005 was on September 22 at 3:23 p.m. However, to be consistent with our observation pattern, the site was observed and recorded on September 21. The azimuth difference between the two days at the time of observation was one tenth of one degree (208.2° on the 21st and 208.1° on the 22nd), while the altitude variance was -.4°.

Figure 37. September 21 observation

As expected, the sun shaft and shadows (Figure 37) closely paralleled the image pattern observed for the vernal equinox on March 21 (Figure 20). This parallel effect strongly suggests that the Sinagua marked the vernal equinox on a day the sun's altitude was at the mid-point of the two observations (Appendix A).

The effect began at approximately 1:27 p.m. and lasted for six minutes. The day was partially overcast which produced softened images for most of this period. The light shaft from between the shadow stones highlighted a major concentric circle image (#1). In addition, a secondary concentric circle image (#2) was tangent to the base stone's upper shadow line (Figure 38). Further, the next major concentric circle image (#4) was also tangent to the bottom shadow line of the base stone.

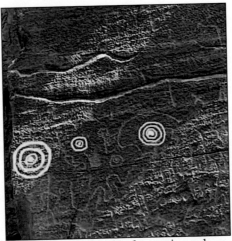

Figure 38. September 21 observation enhanced

As with the March 21 event, the far right concentric circle (#7) appeared to be in line with a shadow line at the base of the image, however, due to the subdued sunlight it was difficult to confirm. Whether this particular glyph is part of the equinox patterns will have to await future observations.

The site was also visited on September 22 and 23 to observe any differences. Unfortunately, the day was almost completely overcast on September 22, preventing detailed recording. September 23, however, was clear but the sun shaft was already moving up the panel and bisecting the major concentric circle (#1) into its second circle, and no longer tangent to the secondary concentric circle glyph.

Late September is the time for any remaining crop harvesting and ceremonies. As with the March 21 event, there are no maize stalks involved. This is consistent with the completion of the main corn harvests in August. However, since corn planting can continue until the summer solstice, late corn can be harvested in late September.

OCTOBER 21

As the sun continues in a southerly direction the light shaft moves higher up the solar panel. There are no images above the major concentric circle glyph designated as #1. The sun shaft does, however, continue to shine on the panel as shown in Figure 39. The sun shaft began at 1:40 p.m. and lasted for seven minutes before beginning to dissipate.

Figure 39. October 21 observation

It was of interest to note that the bottom edge of the light shaft was exactly at a notched area in the panel's edge (Figure 40). The width of the shaft was the width of this notched area, before the panel takes a diagonal direction. This was a similar effect as noted with the notches lower down the panel that came into play at other observations. The notch, upon a closer examination suggests that it may have been worked to form this notch, but this has not been confirmed.

Figure 40. October 21 observation enlarged

If one accepts the agricultural-ceremonial premise of this study, the fact that there are no images higher on the panel that are affected by the sun shaft or shadow lines would not be surprising. This was expected from the examination of historic Pueblo ethnography for this time of year. The harvests are over. Indications are that if ceremonies are performed, they are of lesser importance. In addition, these ceremonies appear to have been dictated by the rise of a new moon rather than by the sun's position.

NOVEMBER 21

As the sun continues on its southerly course, the light shaft moved higher up the solar panel (Figure 41). As pointed out earlier, there are no images in this area. The sun shaft continued to shine on the panel as shown in Figure 42. The sun shaft began at 1:57 p.m. and lasted for about seven minutes before beginning to dissipate.

Unlike the October effect, there was no notch to mark the width of the light shaft. However, there were two interesting features. First, a second light shaft appeared below the base stone. The end of this second light shaft highlights the top of the notch that marked the width of the sun shaft on October 21.

The second feature of interest is near the center of the light shaft. Notice that there is a natural outcropping in the rock face that throws a shadow into the light shaft. This outcropping appears to have been enhanced by chipping away a portion of the rock face (Figure 42) as evidenced by the apparent removal of the surface area with its desert varnish. Further close study is needed to confirm whether this surface has, in fact, been worked.

Figure 41. November 21 observation

Figure 42. November 21 observation enlarged

49

For the historic Pueblo, November marks the start of a new year. The *Wuwutsim* ceremony is observed in November. As with several festivals after the autumnal equinox, there are differing historical reports as to whether this ceremony was a solar or lunar ceremony. One report (Steward 1931) has this ceremony announced during the first quarter moon of November. The moon's orbital plane shifts as compared to the earth's over 18.6 years. Because of this, the dates for observance of any lunar ceremony would necessarily vary over several weeks from year to year.

Others reported that the exact day of the ceremony varied with solar observations. But the recorded dates based on solar observation also varied. For example, it was recorded to have occurred on November 17 in 1892, 1893 and 1898, November 18 in 1921, and November 21 in 1911 (McCluskey 1977).

It is possible that the outcropping may have been a marker to designate the start of the *Wuwutsim* ceremony if the Sinagua observed this festival and used solar observation for its marking. Alternatively the outcropping was a way to mark the passage of another "month."

8

WINTER SOLSTICE SEQUENCE

DECEMBER 21 -- WINTER SOLSTICE

The Winter Solstice usually occurs on December 21st or 22nd
(Appendix B). The Winter Solstice in 2005 at V Bar V was on December
21 at 11:35 a.m. The site was observed and recorded on that date.

It is of interest to note that the winter solstice interactions at V Bar V
are defined by a notch (Figure 44). This notch is a natural crevice between
the solar panel's bluff face and a natural pillar of red sandstone to its west.
As the sun proceeds southward, it is framed by this notch beginning in late
November. The sun appeared at the top of the notch on November 21, as
shown in Figure 45. This photo was taken through a Baader AstroSolar
filter that permits the sun to appear in its natural neutral white color while
darkening the surrounding terrain.

Figure 44. Solar notch

Figure 45. Sun at top of solar notch on November 21

As with any observation of the sun, the position of the observer is important to determine the intended view. It was assumed that the Sinagua observation point was against the face of the solar panel (Figure 46). Since it is known that the surface level at the time was about 90cm (35.4 in) lower than the current ground level, these pictures were taken while standing in a 4-foot open excavation pit at the base of the solar panel to simulate ground level in 1150 A.D.

Figure 46. Author taking winter solstice images

At 1:59 p.m. on December 21 the sun moved to the lowest point in the notch (Figure 47). The sun produced three effects from this position. One of the most striking was the casting of a sun dagger in the shape of the notch onto the ground at the foot of the Solar Panel (Figures 46 and 48). As the sun continued west, the point of the notch was recessed into an open excavation trench at the base of the panel. There are three layered boulders to the north of the panel. The sun dagger eventually stretched onto these boulders (Figure 49). The determination of whether the tip of the dagger falls onto a petroglyph below the surface will have to await possible future excavations.

sun position

Figure 47. December 21

Figure 48. Sun dagger

Figure 49. Sun dagger on boulders

The second feature observed was on the solar panel. A very faint shaft of sunlight was produced between the stones at 2:16 p.m. High altitude clouds moved into the area during the observation and softened the sun's effect, preventing a quality photo of this effect.

As the sun continued to move to the west, past the pillar, the sun began to strike the solar panel again at 3:10 p.m. At the top of the pillar there is a notch that appears to be natural, although we have been unable to climb to that level due to the fragility of the sandstone. As the sun moved past the pillar, sunlight used this notch to cast a "sun wedge" on the lower left of the Solar Panel. The sun wedge eventually elongated to strike the far-right concentric circle (#7) bisecting it in the center (Figure 50). The width of the shaft matches the wedged-shaped element to the right of the concentric circle. This event completed the detailed recording of solar effects on all seven of the concentric circles.

Figure 50. Winter solstice sun wedge on panel

The winter solstice festival is called *Soyal* and it has been recorded as a significant solar ceremony (Stephen 1936; Zeilik 1985; McCluskey 1990). The key purpose of Soyal is to compel the sun to turn back to the north from its southward swing. White (1962 cited in Zeilik 1985) has argued that the summer and winter solstice ceremonies should be referred to as solar rather than solstitial because they occur at times sometimes removed from the astronomical solstice. For example, Stephen (1936) observed Soyal ceremonies occurring on dates ranging from December 20 to December 26. Coincidentally, he noted that Soyal was determined when the Sun set in a particular notch in the San Francisco Mountains. During Soyal, a few Katsinas appear. The appearance of these Katsinas starts the season of the return of all Katsinas in January and February.

JANUARY 21

Observation continued to plot the path of the sun along the solar panel on January 21 (Figure 51). The sun shaft began to appear at 2:20 p.m. As expected, the pattern repeated that noted for November 21 (Figure 42).

For the Hopi, festivals are held among the villages at this time. The name of the festival varies among the villages, but are all held to mark the return of the Katsinas. There are differing accounts as to whether this was a solar or lunar festival. McCluskey (1977) recorded it as a lunar festival determined by the arrival of the New Moon. Alternatively, Zeilik (1985) reports that, in some villages, this was a solar festival. If lunar-based, this ceremony would have been performed at various times in January or early February due to the Moon's varying path from year to year (over an 18.6 year period). The date can also vary if solar-based since they were dependent upon sunrise/sunset positions.

Figure 51. January 21 observation

No significant manifestations appear on the solar panel. As a result, no ethnographic analogy can be made. All that can be stated with assurance is that the passage of another month is recorded on the panel. What, if anything, the Sinagua did at this time is unknown.

FEBRUARY 21

The 12-month observation cycle (started in March 2005) was completed with the solar panel effects noted on February 21 (Figure 52). The sun shaft began to appear at 2:04 p.m. As expected, the pattern repeated that noted for October 21 (Figure 40).

Powamuy, or *Bean Dance* ceremony, occurs in late February or early March, and celebrates the return of the Katsinam to the Hopi villages. Beans, corn and some other seeds are brought into the heated kivas to force plant growth for ceremonial purposes and as omens for the coming planting season.

This festival is also reported as either a lunar or solar festival. Again, no significant manifestations appear on the solar panel. As noted with the January observation, an ethnographic analogy for V Bar V cannot be made. All that can be stated is that the passage of another month is recorded on the panel.

Figure 52. February 21 observation

9

FALLING SPIRAL ELEMENT

Throughout most of the monthly observations, the petroglyph referred to as a "falling spiral" has produced interesting effects. This element (Figure 53) is composed of several features. First, there are two "dots" or markers – one at the top and another near the end. Next, there are seven lines of varying lengths in a stair step pattern. Third, after the lower dot, four short lines in a tight pattern end in a loose spiral in a clockwise pattern.

Figure 53. Falling Spiral element

The sun shaft patterns on the element, as described in the Vernal Equinox and Summer Solstice chapters, presents several plausible interpretations. To review the observed effects:

1. April 21 – The uppermost "dot" is touched by the sun shaft, that then extends to the third intersection of the glyph. The dot and three lines are within the sun shaft.

2. May 21 – The sun shaft now extends over the fifth and sixth segments of the glyph.

3. June 21 – The sun shaft continues down the glyph now extending over the sixth and seventh segments, ending at the lower "dot."

The observations of July and August repeated the effects of May and April, respectively. Clearly, this glyph is tied to the sun shaft images in a very deliberate manner. Since it begins with the period associated with early corn planting, interpretative analysis will start with a review of the agricultural climate at the time.

SACRED MOUNTAIN BASIN CLIMATIC CHANGES

Several climactic changes have been recorded in the American Southwest that corresponded to similar worldwide changes. The most relevant study related to the Sacred Mountain Basin area was by Davis and Shafer (1991). They conducted pollen analysis of the sediment in nearby Montezuma Well. The Well is a limestone pond south of the Sacred Mountain Basin fed by artesian springs that drains through a cave connecting the Well to Beaver Creek. They found that prior to about 6,000 B.C. the Well and Beaver Creek were well-developed riparian vegetation areas. This period was then followed by a period of reduced stream flow, suggesting a reduction in riparian habitat.

After A.D. 450 the water levels began to rise steadily during what is referred to as the Southwest (or Medieval) Warm Period that stretched from about A.D. 900 to 1400. Sediment pollen analysis at the Well and two other locations within the Arizona Monsoonal Boundary (Davis 1994) indicated increased lake levels from A.D. 700-1350. As might be expected of agricultural societies, this time of monsoon-dominated precipitation corresponds with the time when the agriculturally-based prehistoric cultures of the Southwest came into existence, as well as their decline. Each of these groups lived within the Arizona Monsoonal Boundary that was the main source of moisture for their crops. The study also confirmed that the monsoonal summer precipitation was greater during this period than it is today.

The Southwest Warm Period ended by A.D. 1400. This was followed by a period known as the "Little Ice Age." This cooler and drier period lasted until about A.D. 1850 (Peterson 1994). Peterson noted that by the time settlers began to arrive in the area the farming belt had rebounded to conditions similar to those during the time of the prehistoric cultures.

These studies emphasize the importance of the monsoonal rains for the agricultural success of all of the cultures within the Arizona Monsoonal Boundary, including the Southern Sinagua. If, as Peterson states, the current conditions are similar to those of that time, it can be determined from available statistics (for 1896-1992*) that the average start of the monsoon during this period was July 7, with the earliest start on June 16 and the latest on July 25.

---

* http://ag.arizona.edu/maricopa/garden/html/weather/monsoon.htm

FALLING SPIRAL INTERPRETATIONS

With this background, two plausible interpretations are offered for the falling spiral glyph:

Monsoon Prediction: The summer monsoons were vital to the success of the planting season. If plantings were not done in time for sufficient rooting of the plants, heavy monsoonal rains could wash the seedlings away. It was, therefore, important to know when these heavy rains may occur. As described earlier, the monsoonal season at the time of the Sinagua should be similar to the season after 1850. The spiral at the base of the line below the "dot" may symbolize water (Bostwick 2002) "because water eddies sometimes look like they move in a spiral." The dot highlighted on June 21 suggests a notice of the earliest possible start of the monsoons. This date would be consistent with available historical averages.

Planting Periods: It was observed (Forde 1931) that the entire Hopi planting season from the end of April to the summer solstice was "divided into a number of smaller periods of rather less than a week in duration, named from successive horizon points behind which the sun rises. All these points are announced by the Sunwatchers." It is interesting to note that the "dot" at the top of the glyph is highlighted on April 21, the beginning of the planting season, and the lower "dot" is highlighted on June 21, the end of the planting season. This glyph may represent the span of planting times, while the segments represent the smaller planting periods described by Forde.

10

SINAGUA CULTURAL ASTRONOMY

The ultimate goal of cultural astronomy is "to deepen one's understanding of culture by attempting to make their observations of the heavens intelligible in terms that are meaningful in that society" (McCluskey 1996 cited in Aveni 2003). This goal seeks to discover links between astronomical practices and the economic, spiritual, ritual and belief systems of that culture. Based on the observations made during this survey, some such links can be suggested.

ECONOMIC LINK

The Vernal Equinox sequence presented plausible alignments to suggest an economic (agricultural) link by signaling the time for the planting of ceremonial corn (March 21), early corn (April 21) and the main corn crop (May 21). The short growing season and minimal rainfall predisposed the need to pinpoint the times of planting and harvesting. This was a significant motive for this agricultural-based culture to watch the sky and mark the passage of time.

SPIRITUAL LINK

The practical nature of these glyphs in providing agricultural information comes as no surprise. However, this should not limit the consideration of other purposes as well. Some of the earliest studies of the V-V site have suggested that it was a very spiritual place to the Sinagua. The strongest evidence for this has been the presence of 69 glyphs containing cupules, or circular depressions that have been tied to religious ceremonies that are similar to practices performed in historic Pueblo societies today (Cole 1990).

In Chapter 6, the potential solar marking of the start of the Niman ceremonies was described. These ceremonies, marked by prayers, meditation and dance, showed the linkage to the Katsina spirits and their importance in their lives.

RITUAL LINK

The ritual link to the V Bar V is strengthened by the presence of the solar calendar since the sun watchers were also religious leaders. The sun shaft and shadow castings were part of a public display to announce events, hold ceremonies or provide teaching. Since there were no significant dwellings close to the petroglyph site, people came to the site for these specific purposes. Thus, it is reasonable to suggest that the V Bar V petroglyph site was a sacred space set apart from the space of daily living, as a ceremonial center.

Most major calendrical rituals among historic Pueblo societies are based on careful observation of the sun. As shown on the V Bar V solar panel, these observations are based upon sunlight and shadow falling on the concentric circles glyphs, referred to as Sun Father. The Sun Father's habitual motion across the sky validates the cycle of the year that provide order to sacred space as follows:

On the Vernal Equinox, the Sun Father has arrived from his winter house to provide the warmth for the rebirth of the earth and the fertility of the land.

On the Summer Solstice, the Sun Father has arrived at his summer house where he pauses before beginning his return to his winter house. The ceremonies are performed to convince the Sun Father to linger so that frosts do not come too early. (At the solstices, the azimuth of the sun holds nearly steady for four days.)

On the Autumnal Equinox, the Sun Father's passing begins the process of death to be reborn again in the spring.

On the Winter Solstice, the Sun Father has returned to his winter home where he pauses again. Ceremonies are held to compel the Sun Father to return to his summer house so that the cycle of life can continue.

The historic Pueblos believe they are fundamentally dependent upon the Sun Father for whom they honor with shrines. Sun shrines have been recorded at prehistoric Ancestral Puebloan, Mogollon and Hohokam locations. It would therefore be consistent to hold that the V Bar V site was used similarly as a sun shrine.

BELIEF SYSTEM LINK

As a ceremonial center, if V Bar V was a place where time was marked, it may have been by the sun's passing. This was used as a mechanism to unify the people – where the cosmos was employed as a means of validating their rituals, festivals and ceremonies. The completion of one cycle of time was celebrated with the beginning of another. This can be seen today through the monthly Pueblo festivals and ceremonies. The explanation of this cyclical order by the Sun Priest, through religious stories and myths, provided an explanation of the cosmos.

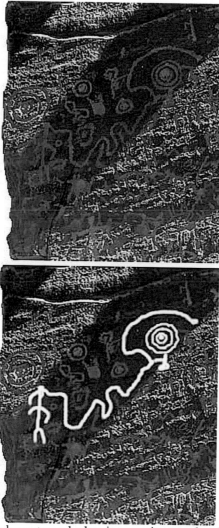

Figure 54. Equinox base stone shadow imagery in natural and enhanced states

During the Vernal Equinox display, an interesting image is highlighted by the base-stone's shadow. Figure 54 shows this image in the natural and enhanced state.

It is believed that rock art elements were probably directly related to ceremonial activities (Weaver 1984). If the V Bar V is a ceremonial center then the presence of ceremonial imagery used to explain the cosmos and the people's relationship to the Sun Father would be expected. It is very curious to note that this elongated imagery, linking several images, is perfectly framed by the base-stone's shadow line only on the equinox (both vernal and autumnal).

Figure 55. Equinox element breakdown

Though highly speculative, this image can be broken down into components (Figure 55) that have been recognized in earlier studies. Segment "b" is reminiscent of kinship or blood lines proposed as a motif used by shaman to show a natural relationship such as offspring or a supernatural relationship (Ritter and Ritter, 1976). Segment "a" is nearly identical to an Ancestral Puebloan image located in the Petrified Forest that was interpreted as a birthing scene (Martynec 1985).

To demonstrate how speculative such an exercise can be one can also suggest that this segment is reminiscent of a snake with an oval head to represent the Snake Clan (Michaelis 1981), or the figure to the left is using a snake whip used by snake priests (Fewkes 1892).

Segment "c" is the familiar Sun Father image (Ellis and Hammack 1968). Segment "d" is much more difficult to associate since the shape's meaning is unclear but it clearly links the Sun Father to the kinship line.

Ritter and Ritter (1976) have suggested that lines that join component parts of a site are a shamanistic device "to indicate a temporal, spatial, natural, or supernatural relationship of the components or to indicate that the panel is to be considered as a composition or whole." This may be the best that can be said about this image. It certainly appears to be storytelling or ceremonial imagery with an overall composition. Further, it is relatively safe to note the fact that the composition is perfectly outlined as part of the public display of the equinoxes. This suggests a cosmological meaning with the cycle of fertility and rebirth in the Spring and death and hope in the Autumn. Additional research and empirical evidence is needed to come to a more definitive interpretation of this imagery.

11

DETERMINING ASTRONOMICAL INTENT

This survey has shown that great care was taken to mark the passage of time and the arrival of specific points in time. This tends to support the primary hypothesis that a cycle of annual, calendrical-based rituals and practices were associated with the Sinagua of the Verde Valley. The following section examines this hypothesis against a set of established criteria for determining purposeful intent.

An attempt to ascribe astronomical intent to a rock art site will always face skepticism. Certainly there are no direct ancestors who can attest to such sites. Ancient astronomers did not have mathematical formulas or almanac references. Their skill was in naked-eye observation and the result of their work needs to be assessed from that perspective.

Archaeologists usually require some degree of precision for acceptance beyond a reasonable doubt. That precision implies astronomical intent in the creation of the site. In order to establish intent at the V Bar V Heritage Site, and to provide a degree of confidence in such intent, an eleven point criteria was applied (McGlone et al. 1999):

1. **Regional context of other alignments** – A single ascribed archaeoastronomy site far removed from similar sites is less convincing than if a site is found in a region of similar confirmed sites. Other archaeoastronomy sites have been ascribed to the Ancestral Puebloan, Hohokam, Mogollon and Sinagua in other parts of northern Arizona. Therefore, the V Bar V site can be said to be a part of a regional context of other alignments.

2. **Deliberate or distinctive sighting point** – This criterion usually applies to sunrise and sunset sites at which the observer must be at

a specific sighting point to make the observations, otherwise the azimuth readings would be incorrect. Other than for the Winter Solstice sun notch observation, there is no such sighting point at V Bar V since it is a midday site.

3. **Inclusion of a definite pattern regarding time to read** – For midday sites, indirect alignment by the maximum or minimum position of shadow or light, is a common pattern. As has been shown at V Bar V, this pattern is present for a limited period of time (6-7 minutes) on a consistent basis.

4. **Use of important calendric days** – Important calendric days include the solstices and equinoxes. Archaeological surveys indicate that the area surrounding the V Bar V site was used extensively for agricultural purposes. The V Bar V site records the equinox and solstice calendric days, as well as others important for an agrarian society.

5. **Ethnographic involvement** – This criterion involves the matching of sites to the cultural values and rituals of the inhabitants. As described with each observation point, specific ritual or agricultural necessity has been presented for most events. As stated earlier, this study focused on Hopi ceremonies since both Hopi and archaeologists recognize Sinagua as being one of the groups that is ancestral to Hopi. Since astronomical associations have been documented at Hopi as indicators for the scheduling of their ceremonies, similar associations were likely present in Sinagua society. Even though many of the Hopi ceremonies are directly related to the Katsina society concept, a world-view concept that does not come into existence until after the demise of the Sinagua as a viable cultural group, it is likely that similar observations and ceremonies took place before the advent of the Katsina concept.

6. **Use of astronomical symbols** – Most rock art sites have a variety of images upon which light and shadow may provide interesting effects. Archaeoastronomy sites must, obviously, employ established astronomical symbols in order to ascribe astronomical meaning. As has been shown, the V Bar V uses astronomical symbols of concentric circles, spirals, snake-like and sun-like glyphs, that have been documented at other astronomical sites.

7. **Uniqueness or definitiveness of construction** – Astronomical sites have a certain uniqueness or creativity that sets them apart. The V Bar V site makes creative use of light and shadow to produce complex geometric matching of multiple symbols. The play of sun and shadow produced by two unusual stone outcroppings is unique. Likewise, the use of multiple symbols for every astronomical event shows a definitiveness of construction. The elements tracked in this study were identified as the only ones that were consistently and completely illuminated by the sun shaft or aligned with the shadow lines. Other elements on the panel that did not fit this criteria were discounted.

8. **Progressive sequencing of a series of lighting events** – While a single event site, such as for the summer solstice, may raise some skepticism, a site with a progressive series of events is much more compelling. As has been shown, the V Bar V site displays a definite sequencing of lighting events for a full 12 months, including the equinox, solstice and intervening calendric markers.

9. **The accuracy of an alignment to the sun's position on certain days** – Accuracy signifies implicit precision in the intent of the maker of an alignment. The V Bar V displays remarkably complex geometric alignment of multiple symbols to the sun's position on specific days of importance. This precision is the strongest evidence for purposeful intent.

10. **Chipping or shaping of shadow-casting surfaces to match targets** – Multiple petroglyphs on the bluff surface have been created to match the shadow or light targets produced by the shadow stones. Whether the shadow stones are natural geologic features or enhanced is of lesser importance to establishing the deliberate creation of the symbols to match the target shadow lines or sun shaft.

11. **Unusual petroglyph placement** – The astronomical symbols at the V Bar V site are unusual in several aspects: (1) the quantity of symbols; (2) the concentration of such images on only one of the thirteen panels; and (3) the range of the images covering a 28.8 m by 28.9 m area. Considering that the images were all created at a height above the stature of the Sinagua (average male of 5'-6"), adds to their unusual placement and intentional purpose.

Based on this eleven point criteria, as well as the direct observations made over the span of a two years, a high degree of confidence can be applied to a finding that the V Bar V Solar Panel represents a calendar of intentional design.

12

SUMMARY AND CONCLUSIONS

This study has documented the recording of time through the play of light and shadow on petroglyph images and by the use of natural or man-made features on the bluff (Figures 54 and 55) over a 12-month period. The eleven-point criterion applied to the site described in the previous section points to its elaborate geometric organization and strongly suggests deliberate intention in their creation and placement.

It has been shown that the light and shadow patterns display their unique configurations at specific times. The events occur on the 21st of each month. The designation of the 21st for dating is, of course, based on the Gregorian calendar, not available to the Sinagua. Without such a device, the Sinagua needed a point of reference to record time. The most logical reference was the solstice and equinox events. The marking of solstice and equinox by many different cultures has been noted in numerous anthropological studies, further confirming the likelihood that the Sinagua also observed them. These occur at regular and repeated intervals, generally on the 21st of the month. It would have been a simple matter to organize the intervening days into equal periods of (30 days), marking each on what we today record as the 21st of each month. The result was the creation of a full 12 month calendar.

In an earlier section, the agricultural nature of the surrounding countryside was discussed. The development of the solar calendar for specific agricultural events is a logical conclusion. The highlighting of the unique glyph called the "home dance glyph" and the spring/fall composition glyph were intriguing and suggested the possible addition of ceremonial events as well. This, however, is much more speculative and will require additional study.

What is more certain is that the V Bar V solar panel was an integral part of the Southern Sinagua complex that linked time with the mythical, ritual and agricultural cycles of the valley's population.

Figure 56. Summary of solar effects between equinoxes

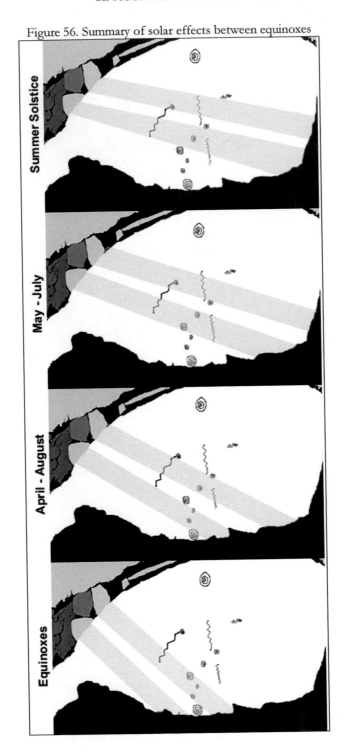

Figure 57. Summary of solar effects around the winter solstice

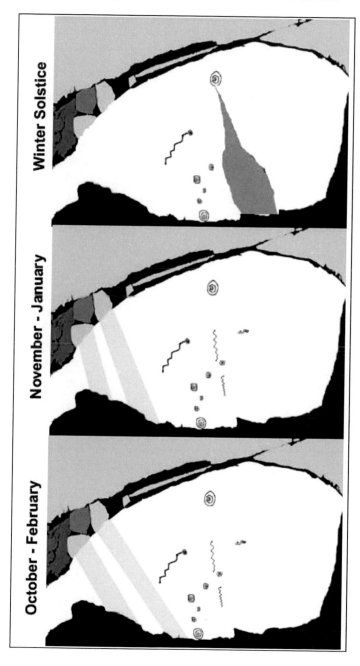

In discussing how the Sinagua might have come upon the knowledge and skill to develop such a calendar, archaeological evidence suggests that calendrical observations have been suggested for virtually every Southwestern cultural group that relied on agriculture. It is also well established in the archaeological record that Mesoamerican (Mexico and Central America) cultures created calendars, probably as early as the 1st or 2nd century A.D. (Kelley and Milone 2005).

Wilcox (1991) has dealt with archaeoastronomy data specifically in the context of Hohokam ceremonial systems. He argues that the structure of Hohokam cosmological beliefs has Mesoamerican parallels. As mentioned in an earlier section there is a ballcourt at the site of Sacred Mountain. The ball game was introduced in Mesoamerica in the 6th century A.D. Cohodas (1975) holds that the ball game was an important ritual held on the equinoxes. The purpose was to ensure the ascent and descent of the sun based on "sympathetic magic." These occasions marked the rebirth and death of the sun-maize deity. The ball courts for these rituals were oriented approximately north south to mark the "southern world" of the winter solstice and the "northern world" of the summer solstice. About a dozen ball courts have been recorded in the Verde Valley, all oriented approximately north south. Their presence strongly suggests that the Sinagua shared the religious symbolism, with its astronomical associations, of Mesoamerica, as did other Southwestern groups.

Wilcox (1999) believes that these ball courts "facilitated ceremonial exchanges" among the different groups. The presence of a solar calendar at the
V Bar V site would have been consistent with this ceremonial exchange and permitted the accurate calculation of the time for the equinox ball games.

Whether this ceremonial relationship existed is speculative. What is more certain is that the V Bar V petroglyph site was an integral part of the Southern Sinagua complex that linked time with the mythical, ritual and agricultural cycles of the valley's population. This yearlong study has suggested several possible astronomically related petroglyphs, and additional study of these photographs may identify other possible associations.

The Sacred Mountain Basin, and particularly the V Bar V Heritage Site, have more interesting stories to tell.

APPENDICES

APPENDIX A: V BAR V HERITAGE SITE SOLAR ALTITUDE AND AZIMUTH
(Green shaded area is Solar Non; yellow shaded area is the time span of the sun shaft)

Standard Time	21-Mar-05		21-Apr-05		21-May-05		21-Jun-05		21-Jul-05		21-Aug-05		22-Sep-05	
	ALT	AZ	ALT	AZ	ALT	AZ	ALT	AZ	ALT	AZ	ALT	AZ	ALT	AZ
12:11 PM	55.4	169.9	67.2	170.8	75.4	168.3	78.1	159.8	74.9	159.6	66.8	168.2	55.4	176.3
12:12 PM	55.5	170.3	67.2	171.5	75.5	169.2	78.2	160.9	74.9	160.4	66.9	168.8	55.4	176.7
12:13 PM	55.5	170.7	67.3	172.1	75.5	170.2	78.3	162.0	75.0	161.3	66.9	169.4	55.4	177.2
12:14 PM	55.5	171.2	67.3	172.7	75.5	171.1	78.3	163.1	75.1	162.2	67.0	170.0	55.4	177.6
12:15 PM	55.6	171.6	67.3	173.3	75.6	172.0	78.5	164.2	75.1	163.1	67.0	170.6	55.4	178.1
12:16 PM	55.6	172.0	67.3	174.0	75.6	173.0	78.5	165.3	75.2	164.0	67.0	171.3	55.4	178.5
12:17 PM	55.6	172.5	67.4	174.6	75.6	173.9	78.6	166.4	75.2	164.8	67.1	171.9	55.4	178.9
12:18 PM	55.7	172.9	67.4	175.2	75.6	174.8	78.6	167.5	75.3	165.7	67.1	172.5	55.4	179.4
12:19 PM	55.7	173.4	67.4	175.9	75.6	175.8	78.6	168.7	75.3	166.7	67.1	173.1	55.4	179.8
12:20 PM	55.7	173.8	67.4	176.5	75.7	176.7	78.6	169.8	75.4	167.6	67.1	173.8	55.4	180.3
12:21 PM	55.7	174.3	67.4	177.1	75.7	177.7	78.7	171.0	75.4	168.5	67.2	174.4	55.4	180.7
12:22 PM	55.8	174.7	67.4	177.8	75.7	178.6	78.7	172.1	75.5	169.4	67.2	175.0	55.4	181.1
12:23 PM	55.8	175.1	67.4	178.4	75.7	179.6	78.7	173.3	75.5	170.3	67.2	175.7	55.4	181.6
12:24 PM	55.8	175.6	67.4	179.1	75.7	180.5	78.7	174.4	75.5	171.2	67.2	176.3	55.4	182.0
12:25 PM	55.8	176.0	67.4	179.7	75.7	181.4	78.8	175.6	75.6	172.2	67.2	176.9	55.4	182.5
12:26 PM	55.8	176.5	67.4	180.3	75.7	182.4	78.8	176.8	75.6	173.1	67.2	177.5	55.4	182.9
12:27 PM	55.8	176.9	67.4	181.0	75.7	183.3	78.8	178.0	75.6	174.1	67.2	178.2	55.4	183.3
12:28 PM	55.8	177.4	67.4	181.6	75.6	184.3	78.8	179.1	75.6	175.0	67.2	178.8	55.3	183.8
12:29 PM	55.8	177.8	67.4	182.2	75.6	185.2	78.8	180.3	75.7	175.9	67.2	179.4	55.3	184.2
12:30 PM	55.9	178.2	67.4	182.9	75.6	186.2	78.8	181.5	75.7	176.9	67.2	180.1	55.3	184.6
12:31 PM	55.9	178.7	67.4	183.5	75.6	187.1	78.8	182.7	75.7	177.8	67.2	180.7	55.3	185.1
12:32 PM	55.9	179.1	67.4	184.1	75.6	188.0	78.8	183.9	75.7	178.8	67.2	181.3	55.3	185.5
12:33 PM	55.9	179.6	67.4	184.8	75.5	189.0	78.8	185.0	75.7	179.7	67.2	182.0	55.3	186.0
12:34 PM	55.9	180.0	67.4	185.4	75.5	189.9	78.7	186.2	75.7	180.7	67.2	182.6	55.2	186.4
12:35 PM	55.9	180.5	67.3	186.0	75.5	190.8	78.7	187.4	75.7	181.6	67.2	183.2	55.2	186.8
12:36 PM	55.9	180.9	67.3	186.7	75.4	191.7	78.7	188.5	75.7	182.6	67.2	183.9	55.2	187.3
12:37 PM	55.9	181.4	67.3	187.3	75.3	192.6	78.6	189.7	75.7	183.5	67.2	184.5	55.2	187.7
12:38 PM	55.9	181.8	67.3	187.9	75.3	193.6	78.6	190.8	75.7	184.5	67.2	185.1	55.1	188.1
12:39 PM	55.9	182.3	67.2	188.6	75.3	194.5	78.6	192.0	75.6	185.4	67.1	185.7	55.1	188.6
12:40 PM	55.8	182.7	67.2	189.2	75.2	195.4	78.5	193.1	75.6	186.3	67.1	186.4	55.1	189.0
12:41 PM	55.8	183.1	67.2	189.8	75.2	196.2	78.5	194.2	75.6	187.3	67.1	187.0	55.0	189.4
12:42 PM	55.8	183.6	67.1	190.4	75.1	197.1	78.4	195.3	75.6	188.2	67.1	187.6	55.0	189.9
12:43 PM	55.8	184.0	67.1	191.1	75.0	198.0	78.4	196.5	75.5	189.1	67.0	188.2	55.0	190.3
12:44 PM	55.8	184.5	67.1	191.7	75.0	198.9	78.3	197.5	75.5	190.1	67.0	188.9	54.9	190.7
12:45 PM	55.8	184.9	67.0	192.3	74.9	199.7	78.2	198.6	75.5	191.0	66.9	189.5	54.8	191.2
12:46 PM	55.8	185.4	67.0	192.9	74.8	200.6	78.2	199.7	75.4	191.9	66.9	190.1	54.8	191.6
12:47 PM	55.7	185.8	66.9	193.5	74.8	201.5	78.1	200.8	75.4	192.8	66.9	190.7	54.8	192.0
12:48 PM	55.7	186.3	66.9	194.1	74.7	202.3	78.0	201.8	75.3	193.7	66.9	191.3	54.8	192.4
12:49 PM	55.7	186.7	66.8	194.7	74.6	203.1	78.0	202.8	75.3	194.6	66.8	192.0	54.7	192.9
12:50 PM	55.7	187.1	66.8	195.3	74.5	204.0	77.9	203.9	75.2	195.5	66.8	192.6	54.7	193.3
12:51 PM	55.6	187.6	66.7	196.0	74.4	204.8	77.8	204.9	75.2	196.4	66.7	193.2	54.6	193.7
12:52 PM	55.6	188.0	66.7	196.6	74.4	205.6	77.7	205.9	75.1	197.3	66.7	193.8	54.6	194.1
12:53 PM	55.6	188.5	66.6	197.1	74.3	206.4	77.6	206.9	75.0	198.2	66.6	194.4	54.5	194.6
12:54 PM	55.6	188.9	66.5	197.7	74.2	207.2	77.5	207.8	75.0	199.0	66.6	195.0	54.5	195.0
12:55 PM	55.5	189.3	66.5	198.3	74.1	208.0	77.4	208.8	74.9	199.9	66.5	195.6	54.4	195.4
12:56 PM	55.5	189.8	66.4	198.9	74.0	208.7	77.3	209.7	74.8	200.8	66.5	196.2	54.4	195.8
12:57 PM	55.5	190.2	66.3	199.5	73.9	209.5	77.2	210.7	74.8	201.6	66.4	196.8	54.3	196.2
12:58 PM	55.4	190.6	66.3	200.1	73.8	210.3	77.1	211.6	74.7	202.5	66.4	197.4	54.2	196.7
12:59 PM	55.4	191.1	66.2	200.7	73.7	211.0	77.0	212.5	74.6	203.3	66.3	198.0	54.2	197.1
1:00 PM	55.3	191.5	66.1	201.3	73.6	211.7	76.9	213.4	74.5	204.1	66.2	198.5	54.1	197.5

Standard	21-Mar-05		21-Apr-05		21-May-05		21-Jun-05		21-Jul-05		21-Aug-05		22-Sep-05	
Time	ALT	AZ	ALT	AZ	ALT	AZ	ALT	AZ	ALT	AZ	ALT	AZ	ALT	AZ
1:01 PM	55.3	191.9	66.1	201.8	73.5	212.5	76.8	214.2	74.4	204.9	66.2	199.1	54.1	197.9
1:02 PM	55.3	192.4	66.0	202.4	73.4	213.2	76.7	215.1	74.3	205.7	66.1	199.7	54.0	198.3
1:03 PM	55.2	192.8	65.9	203.0	73.2	213.9	76.5	215.9	74.3	206.5	66.0	200.3	53.9	198.7
1:04 PM	55.2	193.2	65.8	203.5	73.1	214.6	76.4	216.8	74.2	207.3	65.9	200.9	53.9	199.1
1:05 PM	55.1	193.7	65.7	204.1	73.0	215.3	76.3	217.6	74.1	208.1	65.9	201.4	53.8	199.5
1:06 PM	55.1	194.1	65.7	204.6	72.9	216.0	76.2	218.4	74.0	208.9	65.8	202.0	53.7	199.9
1:07 PM	55.0	194.5	65.6	205.2	72.8	216.7	76.0	219.2	73.9	209.6	65.7	202.6	53.7	200.3
1:08 PM	55.0	194.9	65.5	205.8	72.6	217.3	75.9	220.0	73.8	210.4	65.6	203.1	53.6	200.8
1:09 PM	54.9	195.4	65.4	206.3	72.5	218.0	75.8	220.7	73.7	211.1	65.6	203.7	53.5	201.2
1:10 PM	54.9	195.8	65.3	206.8	72.4	218.7	75.6	221.5	73.6	211.9	65.5	204.2	53.4	201.6
1:11 PM	54.8	196.2	65.2	207.4	72.3	219.3	75.5	222.2	73.4	212.6	65.4	204.8	53.4	202.0
1:12 PM	54.7	196.6	65.1	207.9	72.1	219.9	75.4	222.9	73.3	213.3	65.3	205.3	53.3	202.4
1:13 PM	54.7	197.1	65.0	208.4	72.0	220.6	75.2	223.6	73.2	214.0	65.2	205.9	53.2	202.7
1:14 PM	54.6	197.5	64.9	209.0	71.9	221.2	75.1	224.3	73.1	214.7	65.1	206.4	53.1	203.1
1:15 PM	54.6	197.9	64.8	209.5	71.7	221.8	74.9	225.0	73.0	215.4	65.0	207.0	53.0	203.5
1:16 PM	54.5	198.3	64.7	210.0	71.6	222.4	74.8	225.7	72.9	216.1	64.9	207.5	53.0	203.9
1:17 PM	54.4	198.7	64.6	210.5	71.4	223.0	74.6	226.4	72.7	216.8	64.8	208.0	52.9	204.3
1:18 PM	54.4	199.1	64.5	211.1	71.3	223.6	74.5	227.0	72.6	217.5	64.7	208.5	52.8	204.7
1:19 PM	54.3	199.6	64.4	211.6	71.2	224.2	74.3	227.7	72.5	218.1	64.6	209.1	52.7	205.1
1:20 PM	54.2	200.0	64.3	212.1	71.0	224.7	74.2	228.3	72.4	218.8	64.5	209.6	52.6	205.5
1:21 PM	54.2	200.4	64.2	212.6	70.9	225.3	74.0	228.9	72.2	219.4	64.4	210.1	52.5	205.9
1:22 PM	54.1	200.8	64.1	213.1	70.7	225.8	73.9	229.5	72.1	220.0	64.3	210.6	52.4	206.2
1:23 PM	54.0	201.2	64.0	213.6	70.6	226.4	73.7	230.1	72.0	220.7	64.2	211.1	52.3	206.6
1:24 PM	53.9	201.6	63.8	214.1	70.4	226.9	73.6	230.7	71.8	221.3	64.1	211.6	52.2	207.0
1:25 PM	53.9	202.0	63.7	214.5	70.3	227.5	73.4	231.3	71.7	221.9	64.0	212.1	52.1	207.4
1:26 PM	53.8	202.4	63.6	215.0	70.1	228.0	73.2	231.9	71.6	222.5	63.9	212.6	52.1	207.8
1:27 PM	53.7	202.8	63.5	215.5	70.0	228.5	73.1	232.4	71.4	223.1	63.8	213.1	52.0	208.1
1:28 PM	53.6	203.2	63.4	216.0	69.8	229.0	72.9	233.0	71.3	223.7	63.7	213.6	51.9	208.5
1:29 PM	53.5	203.6	63.2	216.5	69.7	229.5	72.7	233.5	71.1	224.2	63.6	214.1	51.8	208.9
1:30 PM	53.5	204.0	63.1	216.9	69.5	230.0	72.6	234.0	71.0	224.8	63.4	214.6	51.7	209.2
1:31 PM	53.4	204.4	63.0	217.4	69.3	230.5	72.4	234.6	70.8	225.4	63.3	215.1	51.6	209.6
1:32 PM	53.3	204.8	62.9	217.9	69.2	231.0	72.2	235.1	70.7	225.9	63.2	215.5	51.5	210.0
1:33 PM	53.2	205.2	62.7	218.3	69.0	231.5	72.1	235.6	70.5	226.5	63.1	216.0	51.4	210.3
1:34 PM	53.1	205.6	62.6	218.8	68.9	232.0	71.9	236.1	70.4	227.0	63.0	216.5	51.2	210.7
1:35 PM	53.0	206.0	62.5	219.2	68.7	232.4	71.7	236.6	70.2	227.6	62.8	216.9	51.1	211.1
1:36 PM	52.9	206.3	62.4	219.7	68.5	232.9	71.6	237.0	70.1	228.1	62.7	217.4	51.0	211.4
1:37 PM	52.8	206.7	62.2	220.1	68.4	233.3	71.4	237.5	69.9	228.6	62.6	217.9	50.9	211.8
1:38 PM	52.8	207.1	62.1	220.5	68.2	233.8	71.2	238.0	69.8	229.1	62.5	218.3	50.8	212.1
1:39 PM	52.7	207.5	62.0	221.0	68.0	234.2	71.0	238.4	69.6	229.6	62.3	218.8	50.7	212.5
1:40 PM	52.6	207.9	61.8	221.4	67.9	234.7	70.9	238.9	69.5	230.1	62.2	219.2	50.6	212.9
1:41 PM	52.5	208.3	61.7	221.8	67.7	235.1	70.7	239.3	69.3	230.6	62.1	219.6	50.5	213.2
1:42 PM	52.4	208.6	61.6	222.3	67.5	235.5	70.5	239.8	69.2	231.1	61.9	220.1	50.4	213.6
1:43 PM	52.3	209.0	61.4	222.7	67.4	236.0	70.3	240.2	69.0	231.6	61.8	220.5	50.3	213.9
1:44 PM	52.2	209.4	61.3	223.1	67.2	236.4	70.2	240.6	68.8	232.0	61.7	221.0	50.1	214.2
1:45 PM	52.1	209.8	61.1	223.5	67.0	236.8	70.0	241.0	68.7	232.5	61.5	221.4	50.0	214.6
1:46 PM	52.0	210.1	61.0	223.9	66.9	237.2	69.8	241.5	68.5	233.0	61.4	221.8	49.9	214.9

APPENDIX B: DATES AND TIMES OF EQUINOXES AND SOLSTICES A.D. 1000-1200

Vernal Equinox	Summer Solstice	Autumnal Equinox	Winter Solstice
1000-03-20 23:47	1000-06-22 10:27	1000-09-23 13:56	1000-12-21 18:22
1005-03-21 04:51	1005-06-22 15:27	1005-09-23 19:01	1005-12-21 23:30
1010-03-21 09:55	1010-06-22 20:27	1010-09-24 00:05	1010-12-22 04:38
1015-03-21 14:59	1015-06-23 01:26	1015-09-24 05:09	1015-12-22 09:47
1020-03-20 20:04	1020-06-22 06:26	1020-09-23 10:13	1020-12-21 14:55
1025-03-21 01:08	1025-06-22 11:26	1025-09-23 15:17	1025-12-21 20:04
1030-03-21 06:12	1030-06-22 16:26	030-09-23 20:22	1030-12-22 01:12
1035-03-21 11:17	1035-06-22 21:26	1035-09-24 01:26	1035-12-22 06:21
1040-03-20 16:21	1040-06-22 02:26	1040-09-23 06:30	1040-12-21 11:29
1045-03-20 21:25	1045-06-22 07:26	1045-09-23 11:34	1045-12-21 16:37
1050-03-21 02:30	1050-06-22 12:25	1050-09-23 16:38	1050-12-21 21:46
1055-03-21 07:34	1055-06-22 17:25	1055-09-23 21:43	1055-12-22 02:54
1060-03-20 12:38	1060-09-23 02:47	1060-12-21 08:03	1060-06-21 22:25
1065-03-20 17:43	1065-06-22 03:25	1065-09-23 07:51	1065-12-21 13:11
1070-03-20 22:47	1070-06-22 08:25	1070-09-23 12:55	1070-12-21 18:20
1075-03-21 03:51	1075-06-22 13:25	1075-09-23 17:59	1075-12-21 23:28
1080-03-20 08:56	1080-06-21 18:24	1080-09-22 23:03	1080-12-21 04:36
1085-03-20 14:00	1085-06-21 23:24	1085-09-23 04:07	1085-12-21 09:45
1090-03-20 19:04	1090-06-22 04:24	1090-09-23 09:11	1090-12-21 14:53
1095-03-21 00:09	1095-06-22 09:24	1095-09-23 14:15	1095-12-21 20:02
1100-03-21 05:13	1100-06-22 14:24	1100-09-23 19:19	1100-12-22 01:10
1105-03-21 10:17	1105-06-22 19:24	1105-09-24 00:24	1105-12-22 06:18
1110-03-21 15:22	1110-06-23 00:23	1110-09-24 05:28	1110-12-22 11:27
1115-03-21 20:26	1115-06-23 05:23	1115-09-24 10:32	1115-12-22 16:35
1120-03-21 01:30	1120-06-22 10:23	1120-09-23 15:36	1120-12-21 21:43
1125-03-21 06:35	1125-06-22 15:23	1125-09-23 20:40	1125-12-22 02:52
1130-03-21 11:39	1130-06-22 20:23	1130-09-24 01:44	1130-12-22 08:00
1135-03-21 16:44	1135-06-23 01:23	1135-09-24 06:48	1135-12-22 13:09
1140-03-20 21:48	1140-06-22 06:22	1140-09-23 11:52	1140-12-21 18:17
1145-03-21 02:52	1145-06-22 11:22	1145-09-23 16:56	1145-12-21 23:25
1150-03-21 07:57	1150-06-22 16:22	1150-09-23 22:00	1150-12-22 04:34
1155-03-21 13:01	1155-06-22 21:22	1155-09-24 03:04	1155-12-22 09:42
1160-03-20 18:06	1160-06-22 02:22	1160-09-23 08:08	1160-12-21 14:50
1165-03-20 23:10	1165-06-22 07:21	1165-09-23 13:12	1165-12-21 19:59
1170-03-21 04:14	1170-06-22 12:21	1170-09-23 18:16	1170-12-22 01:07
1175-03-21 09:19	1175-06-22 17:21	1175-09-23 23:20	1175-12-22 06:15
1180-03-20 14:23	1180-06-21 22:21	1180-09-23 04:23	1180-12-21 11:24
1185-03-20 19:28	1185-06-22 03:21	1185-09-23 09:27	1185-12-21 16:32
1190-03-21 00:32	1190-06-22 08:20	1190-09-23 14:31	1190-12-21 21:40
1195-03-21 05:37	1195-06-22 13:20	1195-09-23 19:35	1195-12-22 02:49
1200-03-20 10:41	1200-06-21 18:20	1200-09-23 00:39	1200-12-21 07:57

BIBLIOGRAPHY

Aveni, Anthony F.
2003   "Archaeoastronomy in the Ancient Americas," *Journal of Archaeological Research*, 11 (2): 149-191

Bates, Bryan
2005   "A Cultural Interpretation of an Astronomical Calendar (Site #WS 83) at Wupatki National Monument," in Fountain, John W. and Rolf M. Sinclair (eds.), *Current Studies in Archaeoastronomy: Conversations Across Time and Space*, Carolina Academic Press, Durham, North Carolina

Bostwick, Todd W. and Peter Krocek
2002   *Landscape of the Spirits: Hohokam Rock Art at South Mountain Park*, University of Arizona Press, Tucson

Bostwick, Todd W., Paul A Lindberg and Kenneth J. Zoll
2011   "A Geological and Archaeological Study of the Solar Gnomons at the V Bar V Heritage Site," *Verde Valley Archaeology Center Occasional Paper No. 1*, VVAC Press, Camp Verde

Byrkit, J.
1988   "The Pala'tkwapi Trail." *Plateau*, 59 (4).

Cohodas, M.
1975   "The Symbolism and Ritual Function of the Middle Classic Ballgame in Mesoamerica," *American Indian Quarterly*, 2(2): 99-130

Cole, Sally J.
1990   *Legacy on Stone*, Johnson Books, Boulder

Colton, Harold S.
1946   "Fools' Names Like Fools' Faces: Petroglyphs in Northern Arizona," *Plateau*, 19 (1): 1-8

Davis, Owen K. and David S. Shafer
1992   "A Holocene Climatic Record for the Sonoran Desert from Pollen Analysis of Montezuma Well, Arizona, USA," *Palaeogeography, Paleoclimatology, Paleoecology*, 92: 107-119

Davis, Owen
1994   "The Correlation of Summer Precipitation in the Southwestern
        USA with Isotopic Records of Solar Activity During the Medieval
        Warm Period," *Climatic Change*, 26: 271-287

Ellis, Florence H. and Laurens Hammack
1968   "The Inner Sanctum of Feather Cave, a Mogollon Sun and Earth
        Shrine linking Mexico and the Southwest," *American Antiquity*,
        33(1): 25-44

Fewkes, Jesse W.
1892   "A Few Tusayan Pictographs," *American Anthropologist*, 5:5-20.

Fish, Suzanne K. and Paul R. Fish, Eds.
1984   "Agricultural Maximization in the Sacred Mountain Basin,
        Central Arizona," in *Prehistoric Southwestern Agricultural Strategies*,
        Arizona State University Anthropological Research Papers No.
        33, 147-159

Forde, C. Daryll
1931   "Hopi Agriculture and Land Ownership," *Journal of the Royal
        Anthropological Institute of Great Britain and Ireland*, lxi, 357-405

Hays-Gilpin, Kelley with Emory Sekaquaptewa
2006   "Siitalpuva: Through the Land Brightened with Flowers," *Plateau*,
        3(1): 12-24.

Kelley, David H. and Eugene F. Milone
2005   *Exploring Ancient Skies: An Encyclopedic Survey of Archaeoastronomy*,
        Springer Science and Business Media, New York

Kriss, Victor
1989   "Mimbres-Mogollon Archaeoastronomy: Another Connection
        Between Mesoamerica and the Southwest," in Hedges, K. (ed.)
        Rock Art Papers 6, San Diego Museum of Man, San Diego

Mallery, Garrick
1893   "Picture Writing of the American Indians," *Bureau of Ethnology
        Tenth Annual Report 1888-89*, Government Printing Office,
        Washington, D.C

Malotki, Ekkehart
1983   *Hopi Time: A Linguistic Analysis of the Temporal Concepts in the Hopi Language*, Mouton, New York.

Martynec, Richard J.
1985   "An Analysis of Rock Art at Petrified Forest National Park," in Hedges, K. (ed.) *Rock Art Papers 2*, San Diego Museum of Man, San Diego

McCluskey, Stephen C.
2005   "Different Astronomies, Different Cultures and the Question of Cultural Relativism," in Fountain, John W. and Rolf M. Sinclair (eds.), *Current Studies in Archaeoastronomy: Conversations Across Time and Space*, Carolina Academic Press, Durham, North Carolina.
1990   "Calendars and Symbolism: Functions of Observation in Hopi Astronomy," *Archaeoastronomy* (Cambridge), 15: S1-S16.
1977   "The Astronomy of the Hopi Indians," *Journal for the History of Astronomy*, 8: 174-195

McGlone, Bill, Phil P. Leonard and Ted Barker
1999   *Archaeoastronomy of Southeast Colorado and the Oklahoma Panhandle*, Publishers Press, Salt Lake City

McGregor, John C.
1943   "Burial of an Early American Magician," *Proceedings of the American Philosophical Society*, 86(2): 270-298.

Michaelis, Helen
1981   "Willow Springs: A Hopi Petroglyph Site," *Journal of New World Archaeology*, 4(2): 1-23

Patterson, Alex
1992   *A Field Guide to Rock Art Symbols of the Greater Southwest*, Johnson, Boulder

Peterson, Kenneth L.
1994   "A Warm and Wet Little Climatic Optimum and a Cold and Dry Little Ice Age in the Southern Rocky Mountains, USA," *Climatic Change*, 26:243-269

Pilles, Peter J., Jr.
1996a   "By the Banks of Beaver Creek: The V-V Ranch Petroglyph
        Site." Paper presented at the 23rd Annual Meeting of the
        American Rock Art Research Association, El Paso
1996b   "Cultural Affiliation Assessment: Sinagua" in *Cultural Affiliations:
        Prehistoric Cultural Affiliations of Southwestern Indian Tribes*, pp. 189-
        197, USDA Forest Service, Southwestern Region, Albuquerque.
1981   "The Southern Sinagua," *Plateau*, 53(1): 6-17

Preston, Robert and Ann Preston
1987   "Evidence for Calendric Function at 19 Prehistoric Petroglyph
        Sites in Arizona." *Astronomy and Ceremony in the Prehistoric
        Southwest*, Carlson and Judge eds., University of New Mexico
        Press

Ritter, Dale W. and Eric W. Ritter
1976   "The Influence of the Religious Formulator in Rock Art of North
        America," in Bock, R.J., Frank Bock and John Cawley (eds.) *Rock
        Art Papers 3*, San Diego Museum of Man Papers

Sofaer, Anna P. and Rolf M. Sinclair
1983   "Astronomical Markings at Three Sites on Fajada Butte."
        Astronomy and Ceremony in the Prehistoric Southwest, Carlson and
        Judge eds., University of New Mexico Press

Spier, Leslie
1955   "Mohave Culture Items," *Museum of Northern Arizona Bulletin 28*,
        Flagstaff

Snow, Gerald E.
2006   "Petroglyph Calendar Panel at Chavez Pass." Paper presented at
        the 33rd American Rock Art Research Association Annual
        Conference, Bluff, Utah

Stephen, Alexander M.
1936   "Hopi Journal of Alexander M. Stephen," ed. by E.C. Parsons,
        Columbia University Contributions to Anthropology, Number 23,
        Columbia University Press, New York.

Steward, J.H.
1931   "Notes on Hopi Ceremonies in Their Initiatory Form in 1927-
        1928," *American Anthropologist*, 33: 56-79

Tyler, Hamilton A.
1964   *Pueblo Gods and Myths.* University of Oklahoma Press

Weaver, Donald E., Jr.
1984   "Images on Stone: The Prehistoric Rock Art of the Colorado
        Plateau." *Plateau* 55 (2).

Wilcox, David R.
1991   "The Mesoamerican Ballgame in the American Southwest," in
        Scarborough, V.L. and Wilcox, D.R. (eds.), *The Mesoamerican
        Ballgame*, The University of Arizona Press, Tucson, 101-125.
1999   "A Peregrine View of Macroregional Systems in the North
        American Southwest, A.D. 750-1250," in Neitzel, Jill (ed.), Great
        *Towns and Regional Polities in the Prehistoric American Southwest and
        Southeast*, The University of New Mexico Press, Santa Fe, 115-142

Williamson, Ray A.
1984   *Living the Sky: The Cosmos of the American Indian.* University of
        Oklahoma Press, Norman.

Williamson, Ray A. and Mary Jane Young
1979   "An Equinox Sun Petroglyph Panel at Hovenweep National
        Monument," *American Indian Rock Art 5*, American Rock Art
        Research Association, 71-75

Whitley, David S.
2005   *Introduction to Rock Art Research.* Left Coast Press, Walnut Creek,
        California

Yava, Albert
1978   *Big Falling Snow – A Tewa-Hopi Indian's Life and Times and the
        History and Traditions of His People.* Crown Publishers, New York

Young, M. Jane
1986   "The Interrelationship of Rock Art and Astronomical Practice in
        the American Southwest." *Archaeoastronomy*, 10: S43-S58
2005   "Astronomy in Pueblo and Navajo World Views" in Del
        Chamberlain, Von, John B. Carlson and M. Jane Young (eds.),
        *Songs from the Sky: Indigenous Astronomical and Cosmological Traditions
        of the World,*   Ocarina Books, West Sussex, U.K.

Zeilik, Michael

1985a   "The Ethnoastronomy of the Historic Pueblos, I: Calendrical Sun
         Watching," *Archaeoastronomy*, 8: 1-24

1985b   "Sun Shrines and Sun Symbols in the U.S. Southwest,"
         Archaeoastronomy, 9: S86-S96

1986   Response to "An Appraisal of Michael Zeilik's 'A Reassessment
         of the Fajada Butte Solar Marker'," *Archaeoastronomy*, 10: S66-S70

INDEX

sliver stone 25
Snake Clan 65
Snake Dance 22

snake-like glyph 19
solar noon 13
Solar notch 51
Solar Panel 3, 25, 27
solstice 16
     summer 13, 62
     solstice, winter 37, 51, 62
Soyal 54, 55
spirals 19
sun dagger 52
Sun Father 62, 64, 65
Sun Priest 63
sun shrine 22
Sun Watcher 10, 22, 29
sun wedge 54
sympathetic magic 74

T
top stone 19
Tutuventiwngwu 23
Tuzigoot 5

U
U.S. Forest Service 3

W
Water Clan 10, 24, 29
Wet Beaver Creek 5, 6, 7, 20
Wuwutsim 50

Z
zenith 13
Zuni 9, 22

Made in the USA
Lexington, KY
11 November 2018